PERSPECTIVES IN HUMAN REPRODUCTION

E. S. E. HAFEZ
Editor-in-Chief

Volume 5

HUMAN REPRODUCTIVE PHYSIOLOGY

by

E. S. E. HAFEZ
Reproductive Physiology Laboratories
C. S. Mott Center for Human Growth and Development
Wayne State University School of Medicine
Detroit, Michigan

ANN ARBOR SCIENCE
PUBLISHERS INC
P.O. BOX 1425 • ANN ARBOR, MICH. 48106

230 Collingwood, P. O. Box 1425, Ann Arbor, Michigan 48106

Library of Congress Catalog Card No. 78-50311
ISBN 0-250-40244-0

Manufactured in the United States of America

PREFACE

This book is the companion volume for Volume 4 entitled *Scanning Electron Microscopy of Human Reproduction.* The fundamentals of human reproductive physiology are summarized with emphasis on clinical applications to andrology-gynecology. A modern understanding of human reproduction is becoming increasingly important in the management of infertility and contraception.

It is hoped that this succinct account will be of interest to medical students, internists, and family planning personnel. Illustrated with a large number of line drawings, the work is supplemented by summary tables. Uncommon terms are defined in the glossary in order to avoid oversimplification of physiological phenomena. The overburdened and busy person will appreciate the conciseness and the defined objectives given herein.

Thanks are due to Ms. Grace Schassberger, biomedical illustrator at the C. S. Mott Center for Human Growth and Development, for her elegant line drawings. Thanks are also due to all authors, journals and scientific periodicals, and book publishers who gave permission to use or adapt some of the illustrations that were borrowed from the literature. Special thanks are extended to Mrs. Ruth Monroe for her editorial skills.

E.S.E. Hafez
Detroit, Michigan

This Volume is Dedicated to

E. M. W.

CONTENTS

SECTION I
FUNCTIONAL ANATOMY AND ENDOCRINOLOGY

SECTION II
FUNDAMENTALS OF HUMAN REPRODUCTION

SECTION III
REPRODUCTIVE LIFE CYCLE

SECTION I

FUNCTIONAL ANATOMY AND ENDOCRINOLOGY

CHAPTER 1

MALE REPRODUCTIVE ORGANS

Male and female fetuses contain two sets of embryonic reproductive duct systems: Mullerian ducts and Wolffian ducts. The gonads of the male fetus produce an organizing or inductor substance (fetal androgen) which causes the regression of the Mullerian duct system and permits the full development of the Wolffian duct system. In the absence of this inductor substance, the femininity of the fetus persists.

The reproductive organs are associated with the urinary system from which they are partially derived. The male reproductive system consists of the testes, duct system, accessory organs and external genitalia (Table I; Figure 1), and is supplied with numerous blood vessels and nerve endings.

TESTIS

Structure

The testes lie in the scrotum, which is a muscular sac located between the upper thighs and behind the penis. The wrinkled scrotal skin, sparsely covered with pubic hair, plays a major role in thermoregulation of the testes, since optimal sperm production requires temperatures below body temperature. In cold weather, the muscles in the scrotal wall contract to bring the testes closer to the body. The artery supplying the testis is highly convoluted around coils of testicular veins. This anatomical relationship facilitates an effective countercurrent mechanism

Table I

Functions of Male Reproductive Organs

Organs	*Functions*
Testes	Sperm production in semiferous tubules Leydig cells secrete androgens responsible for development of secondary sexual characteristics: changes in larynx causing deep voice, hair growth on face, maxillae and pubis, characteristic male conformation Sperm transport from seminiferous tubules to epididymis
Scrotum	Thermoregulation of testes
Epididymis	Secretion of epididymal fluid for maturation of sperm Storage and biochemical maturation of sperm Resorption of testicular fluid
Excretory Ducts: Vas deferens Ejaculatory ducts	Sperm transport during ejaculation from epididymis to urethra
Accessory Sexual Organs: Ampullary glans Seminal vesicle Prostate Cowper's gland	Secretion of fluids which constitute the seminal plasma of ejaculated semen Source of energy and nutrients for sperm Maintain optimal pH for sperm survival Contain biochemical substances needed for sperm motility, *e.g.*, fructose
Penis	Sensory and copulatory organ to deposit semen into vagina Pathway of urine

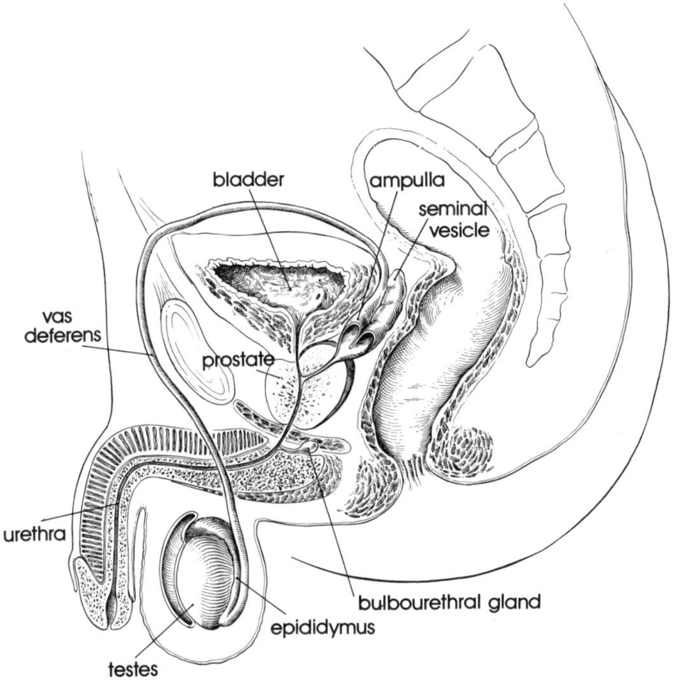

Figure 1. The reproductive organs of adult men.

by which arterial blood supplying the testis is cooled by the venous blood leaving the testis.

The oval-shaped testis (24 × 24 × 50 mm) is divided into some 250 compartments, each containing a few seminiferous tubules where sperm are formed.

Seminiferous Tubules

The highly coiled seminiferous tubules, which make up 80% of the mass of the testis, are arranged in triangular compartments divided by membranes. The tubules, made up of various layers of spermatogenic cells at different degrees of maturation, are surrounded by a "peritubular membrane." With certain testicular disorders (*e.g.*, Klinefelter's syndrome), the peritubular membrane undergoes considerable thickening and hyalinization.

At birth, the seminiferous tubules are not differentiated, but with the onset of puberty, various stages of spermatogenic cells appear, and sperm can be observed within a well-defined lumen of the tubule. In adult men, the epithelium of the seminiferous tubules consists of various stages of spermatogenic cells and supporting Sertoli cells. The spermatogenic cells are arranged in layers of definite sequence as follows: spermatogonium, primary spermatocyte, secondary spermatocyte, spermatid and sperm (spermatozoa).

The seminiferous tubule is surrounded by myoid (smooth muscle) cells which contract to facilitate the release of immotile sperm to the efferent ducts of the testis. The seminiferous tubules connect to a network of some 12 thin tubules—the "rete testis"—which collect the testicular fluid and sperm.

Sertoli Cells

Sertoli cells are large, triangular cells found next to the basement membrane of the seminiferous epithelium. Sertoli cells have an irregularly shaped nucleus and cytoplasm. The cytoplasm has elaborate thin processes which surround several adjacent spermatogenic cells. Thus, Sertoli cells have an irregular shape due to the changing conformation of the various spermatogenic cells embedded within their cytoplasm. After the release of sperm from their cytoplasm, Sertoli cells are reduced in size. Sertoli cells perform several functions:

Coordination of spermatogenesis. Sertoli cells provide mechanical support and metabolic interchange for spermatogenic cells before their release into the lumen of seminiferous tubules; however, the physiological significance of Sertoli cells in spermatogenesis is not fully known.

Endocrine function. Sertoli cells secrete androgens, particularly testosterone and certain hormone derivatives of pregnane and pregnene. It has not been confirmed whether Sertoli cells secrete estrogen or the so-called hormone "inhibin." These cells may also act as an intermediate target system for gonadotropins.

Testicular fluid. Sertoli cells facilitate the formation of testicular fluid required for the transport of sperm from the testes to the epididymis.

Phagocytosis. Lipid-rich residual bodies produced during the transformation of the spermatid into sperm undergo phagocytosis by the Sertoli cells. These residual bodies may provide substrate for the synthesis of steroids. Under certain conditions, Sertoli cells reabsorb degenerating spermatogenic cells.

Leydig Cells

Leydig cells are scattered among the seminiferous tubules forming piles, cords or clusters around the blood capillaries with which they are closely related. Leydig cells undergo morphological and probably maturational changes. These cells are the primary source of testosterone which, supplemented by testosterone produced by Sertoli cells, controls the functional activity of accessory sex organs and the development of secondary sexual characteristics.

Spermatogenesis

Sperm production (spermatogenesis) can be divided in the testis into three principal phases: spermatocytogenesis, meiosis, and spermiogenesis.

Spermatocytogenesis

The spermatogonium is the first stage of germinal epithelium within the seminiferous tubules. Spermatogonia proliferate, giving rise to new stem cells to replace those that have differentiated into spermatocytes. Spermatogonial stem cells are of two types: those that serve as an emergency source of new stem cells ("reserve" stem cells), and those that may begin the process of differentiation into spermatocytes ("renewing" stem cells).

Meiosis

Spermatocytes undergo two maturational divisions to reduce their chromosomal number. In the first maturational division, the primary spermatocyte divides to form two secondary spermatocytes, which undergo a second maturational division to become spermatids. As meiosis progresses, the primary spermatocytes become larger as they migrate from the basement membrane toward the center of the seminiferous tubule. Clusters of spermatids are formed near the lumen of the tubule.

Spermiogenesis

During this last "differentiation" phase, the spermatids do not divide, but undergo remarkable transformation into sperm. Spermatogenic cells undergo three developmental processes: acrosome formation, nuclear condensation and tail formation. Thus, spermiogenesis is a process of cell differentiation rather than of cell division. There is a loss of cytoplasm and a condensation of nuclear chromatin in the spermatid. In the later stages of spermiogenesis, the nucleus flattens, a flagellum appears and a midpiece is formed so that finally a mature sperm is produced. A small portion of cytoplasm, the cytoplasmic droplet, remains attached to the midpiece of sperm.

The mitotic division of spermatogonia into primary spermatocytes is not hormone-dependent, whereas the reduction divisions of primary spermatocyte to secondary spermatocytes and secondary spermatocytes to spermatids seem to be testosterone-dependent. Sperm production is affected by several factors: general health, age, frequency of ejaculation, season of the year, temperature, photoperiod, nutrition, stress and toxic substances.

Spermiation and Sperm Transport in Male Tract

Spermiation, the release of immotile sperm into the lumen of seminiferous tubules, is under endocrine control. Upon their release, sperm are transported into the duct system, facilitated by several physiological mechanisms, *e.g.*, currents of testicular fluid, cilia beat and muscular activity. Most of the testicular fluid in which sperm are transported is reabsorbed into the epididymis. The testicular fluid varies in biochemical composition from blood plasma and lymph. An effective blood-testis barrier separates the seminiferous epithelium from the general circulation.

The Scrotum

The testes descend in the inguinal canal by the seventh month of gestation and emerge into the scrotum shortly before birth. In approximately 10% of male babies, the testes are undescended at birth, but this percentage decreases to 2 to 3% by the end of the first year of life. Failure of the testes to descend results in cryptorchidism.

The scrotum is a thin pouch of skin, muscle, and elastic and collagenous connective tissue. A medial ventral ridge, the scrotal raphe, marks the fusion line of the embryonic genital swelling. The dartos tunic, beneath the skin, is made of smooth muscle fibers and connective tissue. The two tunics, which surround both testes individually, fuse to form an internal scrotal partition.

For optimal temperature for spermatogenesis, the testes must be maintained at a temperature lower than that of the body. Anatomical features of the scrotum and testis provide for the regulation of testicular temperature. The scrotal skin lacks subcutaneous fat, and is very richly endowed with sweat glands. The contraction and relaxation of the scrotal musculature and dartos muscles alter the thickness of the scrotum and its surface area, and vary the closeness of contact of the testes with the body.

The testicular artery is a convoluted structure in the form of a cone, the base of which rests on the testis. The "parmpiniform plexus," which is formed by a network of veins as they leave the posterior border of the testis and pass up the spermatic cord, surrounds the convoluted testicular artery and serves as a thermoregulatory mechanism. This anatomical relationship provides an effective countercurrent mechanism by which arterial blood entering the testis is cooled by the venous blood leaving the testis. The position of the arteries and veins close to the surface of the testis tends to increase direct thermal loss from the testis.

DUCT SYSTEM

The duct system consists of the rete testis, efferent ductules, epididymis, vas deferens and ejaculatory ducts. The seminiferous tubules connect to the rete testis, which is made of cavernous spaces. Multiple efferent ductules connect the rete testis to a single, long, tortuous duct–the epididymis.

Efferent Ductules

The efferent ductules originate in the rete testis and become enlarged upon entering the epididymis. The efferent ductules are lined with nonciliated cells and ciliated cells equipped with kinocilia (motile), whereas the epididymis is lined with columnar sterociliated (nonmotile) and nonciliated cells. The nonciliated have both secretory and absorptive functions. The beat of kinocilia facilitates the transport of sperm from the testes to the epididymis.

Epididymis

Structure

The epididymis consists of three distinct anatomical regions. The *caput*, flattened toward the anterior pole of the testis, connects to the efferent ductules. The *caput* tapers into the narrow *corpus*. The *cauda*, loosely attached near the posterior pole of the testis, merges into the vas deferens (Figure 2). The epididymal duct is surrounded by a firm, connective tissue sheath.

Eight segments can be recognized in the human epididymis as judged by the diameter of the lumen, the height and cytological characteristics of epithelial cells, cell organelles and cell surface. The lumen is distended in the caput region, and is very narrow at the transition of the corpus and cauda. The lining epithelium of the epididymis is made of small angular cells and taller columnar cells with a tuft of stereocilia.

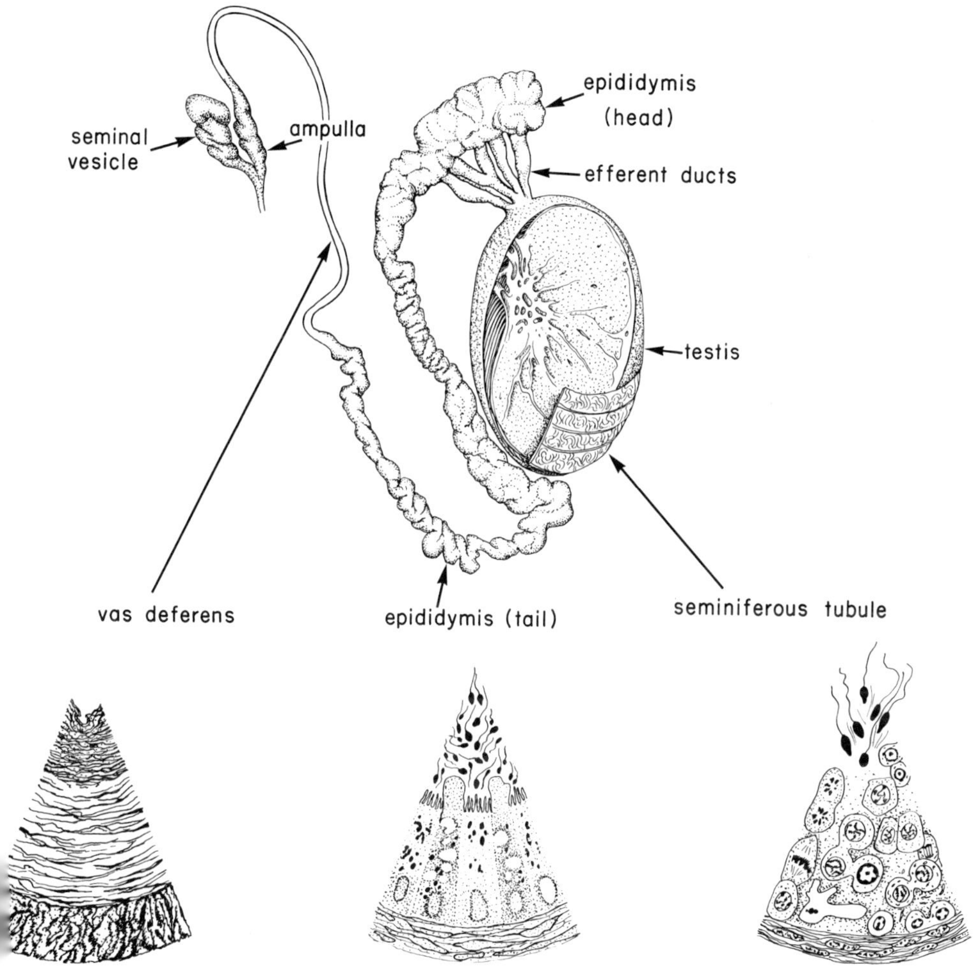

Figure 2. Morphological and histological characteristics of the seminiferous tubules of the testes, the epididymis and vas deferens. Note the differentiation of germinal epithelium and presence of sperm in the seminiferous tubules. The epididymis is lined up by secretory and stereociliated cells, whereas the vas deferens is characterized by its thick muscular wall, which undergoes contraction during ejaculation.

Functions

The epididymal epithelium has both secretory and absorptive functions. It secretes the complex epididymal plasma in which sperm are suspended and undergo maturation. The epididymal fluid may contribute to the constituents of the ejaculated semen. The epididymis plays a role in the storage, transport and maturation of sperm.

Sperm storage. Sperm are primarily stored in the cauda of the epididymis and in the vas deferens. Nonmotile sperm are capable of survival in the epididymal fluid for long periods.

Sperm transport. Transport of sperm through the epididymis takes about three weeks, and is mainly the result of muscular contraction of the epididymal wall.

Sperm maturation. While transported through the epididymis, sperm undergo maturation and acqure motility and fertilizing capacity. Sperm maturation involves morphological, histochemical, physiological, biochemical, biophysical and metabolic changes. The most striking morphological change involves the cytoplasmic droplet, a remnant of spermatid cytoplasm, which migrates caudally from the neck of the sperm to the end of the midpiece. Intrinsic factors, such as aging of the sperm in the epididymis, are not sufficient for the maturational process. Thus, the epididymal environment is needed for the sperm to become fertilized.

Epididymal sperm are immotile and their metabolic activity is low. Their quiescence may be due to several factors, *e.g.*, low partial pressure of oxygen, high level of bicarbonate, mechanical overcrowding, deficiency of fructose for glycolysis, high ratio of potassium to sodium and possible presence of metabolic inhibitors.

Reabsorption. The epididymis is the site of reabsorption of testicular fluid, aged sperm and products of sperm breakdown. Depending on the frequency of ejaculation, there is an equilibrium between breakdown of aged sperm and fullness of the epididymis.

Vas Deferens

The vas deferens is an excretory duct of the testis which extends from the tail of the epididymis. It is highly convoluted and runs parallel to the corpus epididymis; then it gradually straightens, passes through the inguinal canal, loops around the ureter and opens into the urethra. After joining the spermatic cord, the vas deferens enters the pelvic cavity and fuses with the neck of the seminal vesicle to form the ejaculatory duct. The terminal segment of the vas deferens, the "ampulla," is dilated.

The walls of the vas deferens contain a thick layer of smooth muscle which contracts during ejaculation to facilitate the emission of sperm into the urethra.

ACCESSORY SEXUAL GLANDS

The accessory sexual glands are tubular or alveolar structures with secretory and/or absorptive functions. These glands are made of secretory epithelium, connective tissue and smooth muscle fibers. The ampullary glands and seminal vesicles are connected by the vas deferens, whereas the prostate and bulbourethral glands are connected by the uregenital sinus (Figure 3).

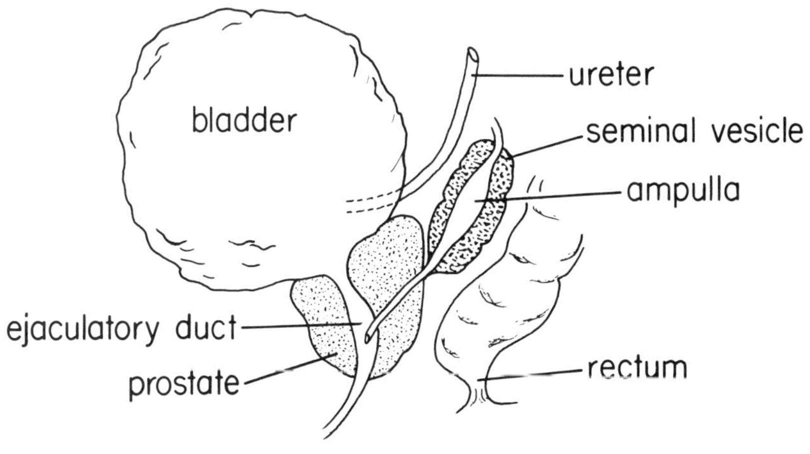

Figure 3. Anatomical relationship of the ureter, bladder, seminal vesicle, prostate and ejaculatory duct (redrawn from Becker, R. F., Ph.D. Thesis, Michigan State University, East Lansing, 1967).

The differentiation, maintenance and functional integrity of these glands depend on androgens. Thus, castration causes atrophy of the epithelium and loss of secretory activity, whereas the administration of testosterone will restore the functional activity and size of these structures.

Secretions of the accessory sexual glands contribute to the complex constituents of the seminal plasma. These secretions provide energy sources for ejaculated sperm and buffer them from pH changes, and provide a liquid vehicle for sperm transport.

Ampullary Glands

Ampullary glands–tubules enclosed by a capsule of connective tissue–are formed from the vas deferens near its junction with the urethra. The irregularly oriented mucosal folds of the ampullary gland show multiple branching. Some of the ejaculated sperm are stored in the ampulla. Following bilateral vasectomy, large numbers of sperm may be found in several subsequent ejaculates, the number of which varies among individual men.

Seminal Vesicles

The two seminal vesicles are elongated and lobulated hollow evaginations of the vas deferens. Each vesicle, 5 to 10 cm long, consists of a tube coiled on itself, giving rise to an irregular diverticulum. The duct of the seminal vesicle joins the ampulla at the ejaculatory duct within the prostate. The ejaculatory ducts from both the left and the right side join the single urethra, still within the prostate.

The lumen of the glands is lined with secretory epithelium. The secretion of the seminal vesicles, released just before ejaculation, is rich in both citric acid and fructose, which provide nutrients to the ejaculated sperm.

The Prostate

The prostate is situated below the neck of the bladder and surrounds the posterior portion of the urethra. It can be palpated during a rectal examination and is the size of a chestnut. The prostate consists of one cranial and two lateral lobes made of

30 to 50 tubuloalveolar structures opening directly into the prostatic urethra through 15 to 30 ducts. Two ejaculatory ducts penetrate the posterior surface of the prostate close to its upper border.

Prostatic fluid, secreted at the rate of 1 to 2 ml/day, is released into the urethra only a few seconds before ejaculation. Secretions from the prostate make up most of the seminal plasma of the semen. One remarkable characteristic of the prostate is the high concentration of zinc. The depletion of zinc in the body following an intravenous injection of diphenylthicarbazone results in severe damage to the prostate. The mode of action of zinc in prostatic physiology is not understood.

Condensation of prostatic secretions may occur within the lumen of the gland of men more than 50 years old, causing overgrowth of the prostatic tissue. Prostatic enlargement is associated with a sequence of events: (1) the urethra encircled by the prostate is narrowed, (2) the outflow of urine is impeded, and (3) the bladder is swollen and its walls become thin.

Bulbourethral (Cowper's) Glands

These are two yellowish-brown, pea-sized glands which join the urethra as it leaves the prostate. The excretory duct of each gland, 3 to 4 cm long, runs parallel to the urethra into which it opens. Bulbourethral glands secrete a few drops of lubricating fluid during sexual stimulation.

Other Accessory Glands

The urethral (Littre's) glands are multiple mucosal glands which open into the cavernous portion of the urethra. Some of these glands are simple outpocketings of the urethral mucosa. Other urethral glands are somewhat larger and are connected to the urethra by a short duct. Several preputial (Tyson's) glands are distributed all over the prepuce and on the neck of the penis.

OTHER ORGANS

The Penis

The penis, primarily made up of spongy erectile tissue, consists of three longitudinal columns as follows: a pair of "corpora cavernosa" forming the bulk of the penis; and a smaller column, "corpus spongiosum," which runs parallel and ventral to the other two, and is traversed by the urethra (Figure 4).

At the tip of the penis, the corpus spongiosum enlarges and extends toward the dorsal surface to form the glans penis, which covers the blunt ends of the corpora cavernosa. The penis differs from its female homologue, the clitoris, only by virtue of its larger size and its possession of the urethra encased in the corpus spongiosum.

A network of large, vascular lacunae, separated by fibrous trabeculae, occupies the three longitudinal columns. During sexual stimulation, the lacunae are filled with blood, the spongy erectile tissue becomes hard and swollen, and the penis undergoes erection. In adult men, the nonerect flaccid penis is about 8 to 11 cm long; when erect, it may be 15 to 20 cm long and 3.5 cm in diameter. Since the clitoris and not the vagina is the center of sexual sensitivity in the female, the length or diameter of the erect penis has little effect on the woman's pleasure during sexual intercourse.

A free fold of skin, the "foreskin," covers the glans. It is common to remove the foreskin surgically within the first week after birth. This circumcision prevents the accumulation of smegma, a waxy secretion secreted by glands beneath the foreskin. Uncircumcised men must pull back the foreskin and wash away accumulated smegma regularly.

Urethra

The urethra extends from the bladder, through the prostate, beneath the pubic bone and through the penis. The external opening at the tip of the penis is called the "urethral or urinary meatus." During urination, urine flows from the bladder through the urethra. During ejaculation, a sphincter muscle at the opening of the bladder closes, so that seminal fluid from the prostate and

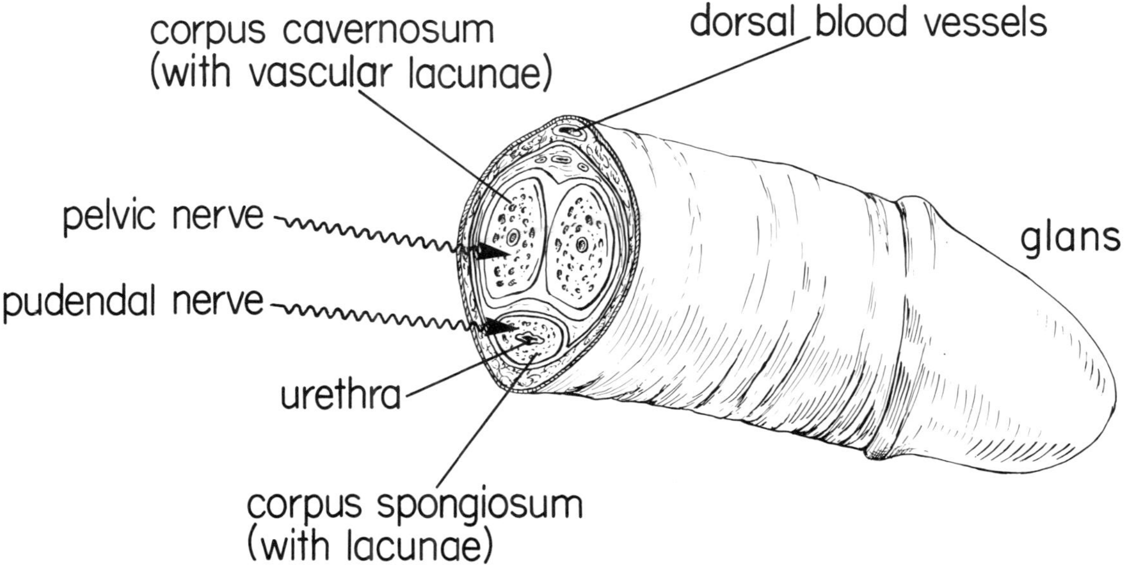

Figure 4. Cross section in penis showing major innervation from S_2-S_4. During penile erection, the lacunae in the corpus cavernosum and corpus spongiosum are filled with blood.

ejaculatory ducts can pass through the urethra without contamination by urine. Thus, it is impossible to urinate during ejaculation. The epithelial lining of the urethra is susceptible to infection by bacteria and viruses, which can enter through the urinary meatus.

Bladder

The bladder is a muscular sac lying behind the pubic bone and in front of the rectum and above the prostate. The bladder collects urine, which flows into it from the kidneys by way of the two ureters.

CHAPTER 2

FEMALE REPRODUCTIVE ORGANS

The female reproductive organs consist of: (1) ovaries; (2) internal reproductive tract-oviduct, uterus, cervix, vagina and Bartholin's gland; (3) external genitalia-mons veneris, labia majora, labia minora, clitoris, vestibule of the vagina and hymen; and (4) mammary glands. The reproductive organs of mature women are illustrated in Figure 5 and their functional anatomy summarized in Table II.

The internal genital organs are supported by the broad ligament, and comprise the mesovarium, which supports the uterus. The reproductive tract is lined by different types of epithelia. The vagina is covered by pseudostratified epithelium. The rest of the tract is covered by one layer of columnar epithelium, basically made up of two cell types: secretory and ciliated (Figures 6 and 7).

The ciliated cells are covered with kinocilia, which beat from the ovarian side to the uterine side, and from the uterine side to the vaginal side (*i.e.*, they beat toward the outside of the body). The secretory cells are covered with numerous microvilli and secrete luminal fluids necessary for the transport and maturation of gametes (sperm and egg) and nutrition of the embryo before it implants in the uterus. The rate of secretions of luminal fluid in any given segment of the reproductive tract depends on: (1) percentage of secretory cells; (2) secretory activity of the cells; and (3) responsiveness of secretory cells to circulating steroid hormones. The structure and ultrastructure of the reproductive tract vary with the age, parity, disease, and stage of the reproductive cycle.

Table II

Structure and Function of Female Reproductive Organs

Organ	*Structure*	*Function*
Ovary	Connective tissue contains ovarian follicles and corpora lutea at various stages of development and smooth muscle fibers	Releases eggs Ovarian follicle secretes estrogen which acts on the female reproductive tract Corpus lutuem secretes progesterone and other hormones which act on the oviduct, endometrium and cervix
Oviduct	Primary and secondary mucosal folds made up of secretory and ciliated cells	Fimbriae picks up egg from ovarian surface Oviductal fluid provides nutrients to sperm and egg Site of fertilization Retains fertilized egg for three days until endometrium is prepared for implantation Transports egg to uterus
Uterus Endometrium	Endometrial glands distributed through stroma	Secretes endometrial fluids necessary for sperm capacitation and nourishment of early blastocyst Site of implantation and attachment of the placenta Maintenance of gestation

Myometrium	Longitudinal and circular muscle cells in spiral arrangement	Contracts to facilitate sperm transport from site of ejaculation to site of fertilization Expulsion of fetus during parturition
Cervix	Complex cervical crypts lined up with mucin-secreting and ciliated cells	Secretes cervical mucus with unique rheological characteristics which restrict sperm migration to uterus except during midcycle Sperm reservoir
Vagina	Pseudostratified epithelium covers vaginal rugae	Copulatory organ Excretion of menstrual blood and vaginal discharges Birth canal
Upper vagina	Wide, distensible, to hold cervical mucus, vaginal fluid and semen	Facilitates sperm migration to external os of cervix
Midvagina	Distensible, but fixed by levator ani muscle	Retains semen pool
Lower vagina	Made up of elastic tissue and hymen or its remnants; partially closed by perineal musculature	Sensory stimuli during precoital excitement Protects against invasion of microorganisms Provides transudates which lubricate vagina during intercourse

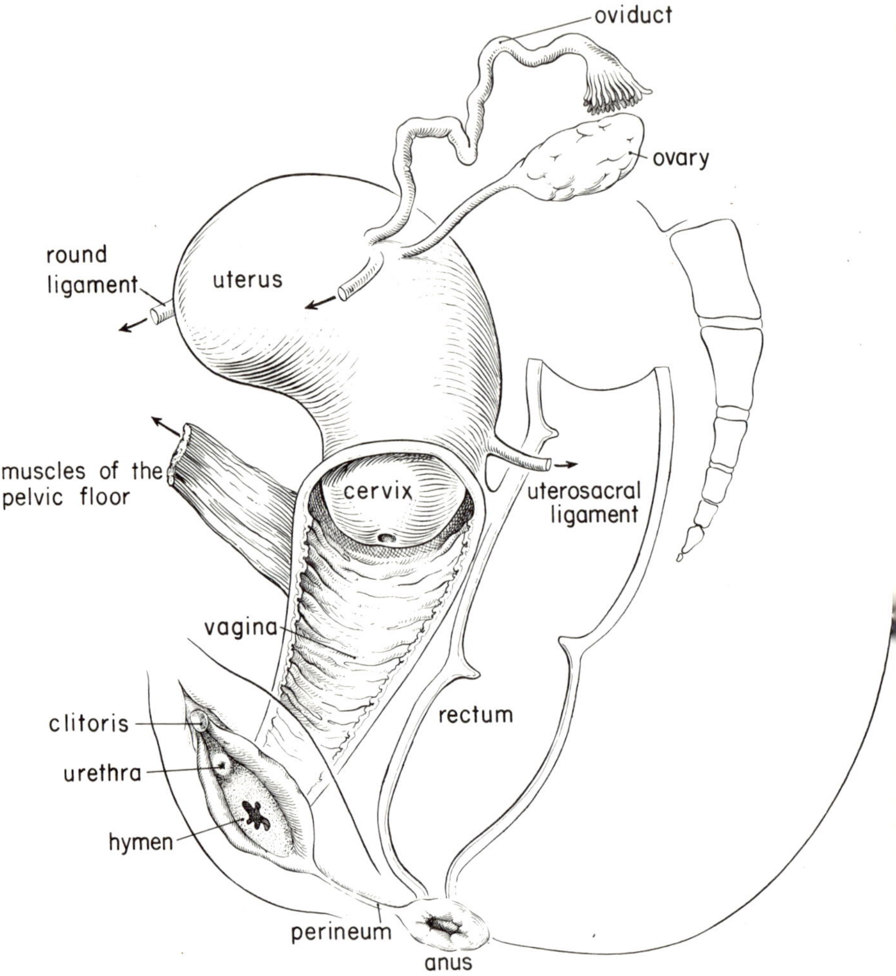

Figure 5. The reproductive organs of the adult woman. The various ligaments support the organs within the pelvic cavity.

OVARY

Unlike the testis, the ovaries remain in the abdomen, and at birth contain a constant number of oocytes. The ovary is attached to a ligament, the mesovarium, which contracts vigorously during the ovulation period causing closer contact between the ovary and the opening of the oviduct. The part of the ovary

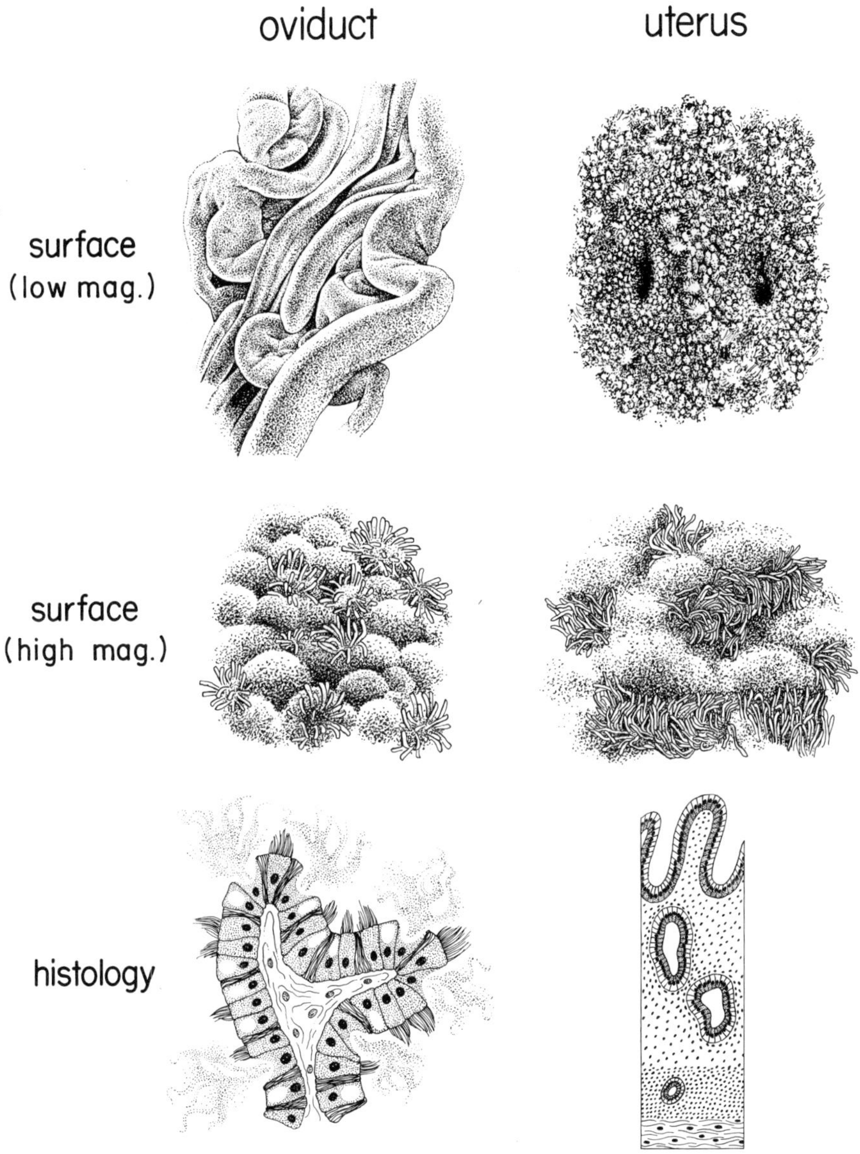

Figure 6. Surface ultrastructure as observed by scanning electron microscopy at low and high magnification (X), and histological characteristics of the oviduct and uterus. Note the complex mucosal folds, and the secretory and ciliated cells in the oviduct. The endometrium is characterized by the presence of endometrial glands, which open in the uterine lumen, and the abundance of secretory cells, which secrete endometrial fluid.

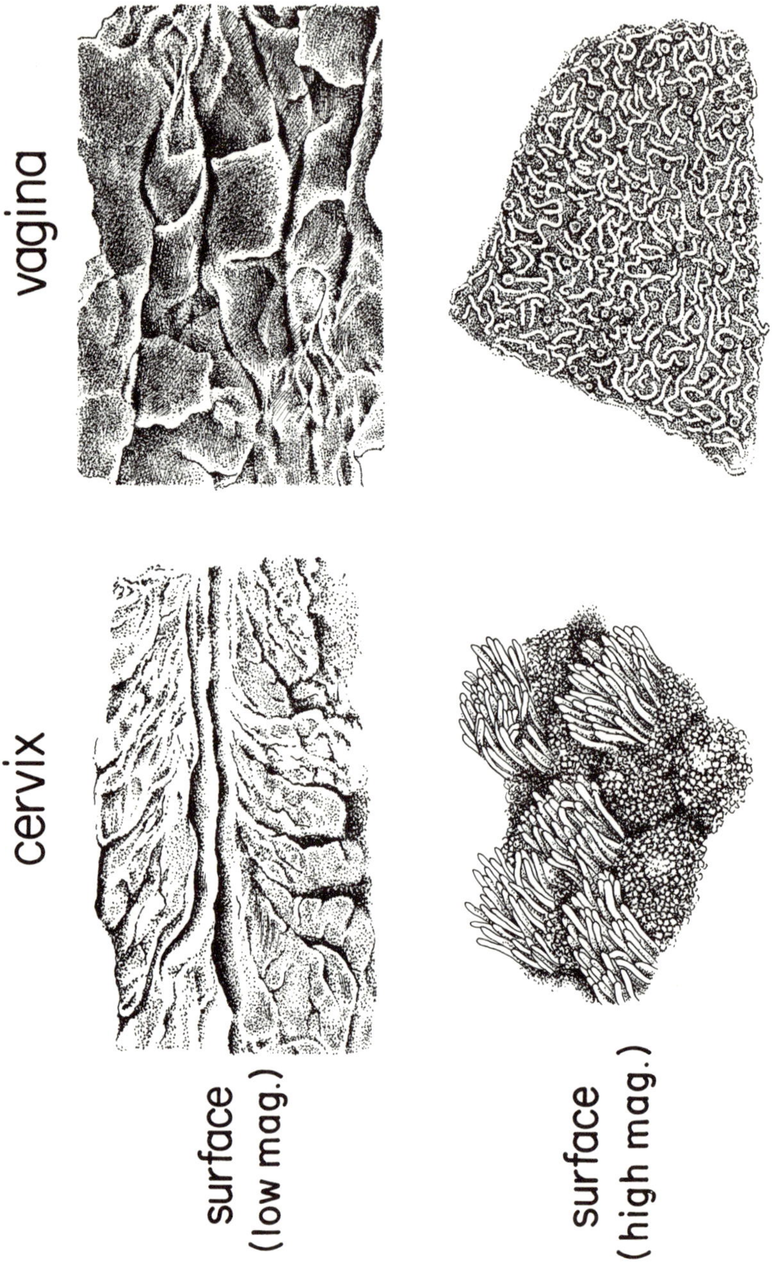
vagina
cervix
surface (low mag.)
surface (high mag.)

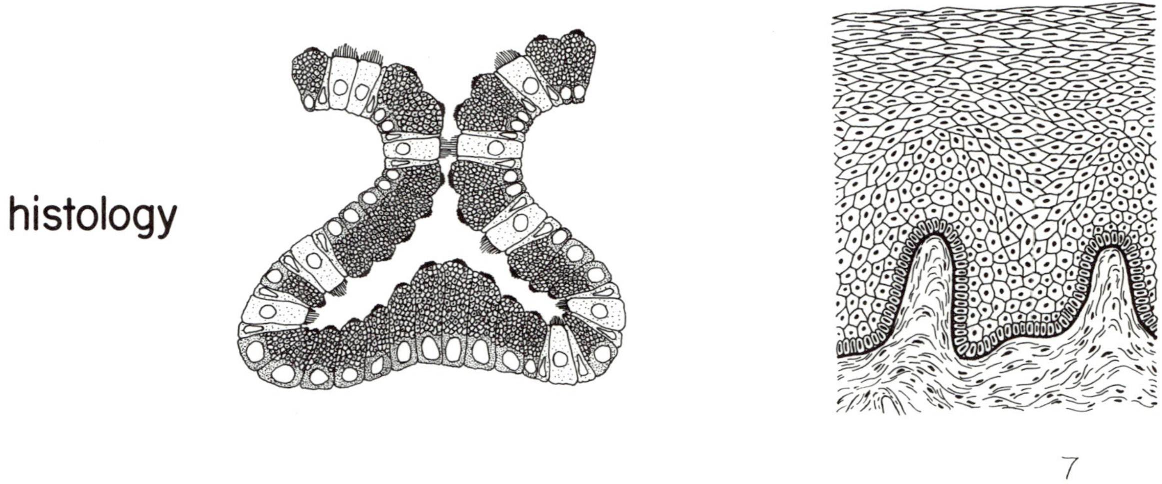

Figure 7. Surface ultrastructure as observed by scanning electron microscopy at low and high magnification (X), and histological characteristics of the cervix and vagina. Note the mucin-secreting cells in the cervical crypts, and the rugae and exfoliated cells on the surface of pseudostratified epithelium of the vagina.

which is not attached to the mesovarium is exposed and bulges into the abdominal cavity. The ovary performs two functions: exocrine–production and release of ova (eggs); and endocrine–secretion of estrogen and progesterone.

The ovary is composed of a medulla and cortex and is surrounded by the germinal epithelium. The medulla consists of irregularly arranged connective tissue containing lymphatic vessels, nerves and smooth muscle fibers. The cortex contains the ovarian follicles and corpora lutea at various stages of development or regression (Figure 8).

Oogenesis

At birth, the ovaries contain several hundred thousand potential eggs. The egg stores in its protoplasm the genetic memory of all previous past generations. Of this large number of potential eggs, only a very small number (400 to 500) successfully complete the maturational changes leading to ovulation. Only a small portion of ovulated eggs undergo subsequent fertilization. Thus, the eggs have the potential to undergo differentiation and maturation over a long period of time from the third week of embryonic development to the premenopausal stage.

Maturation of Ovarian Follicle

The ovarian follicle undergoes several maturational changes which involve an increase in the number of follicular cells around the egg and an increase in the diameter of the follicle itself with subsequent accumulation of follicular fluid (Figure 8).

Atresia

There is a continuous wave of development and atresia (degeneration) of follicles and their oocytes. Only one follicle reaches full maturity from several developing follicles. The endocrine mechanisms of this phenomenon are not fully understood.

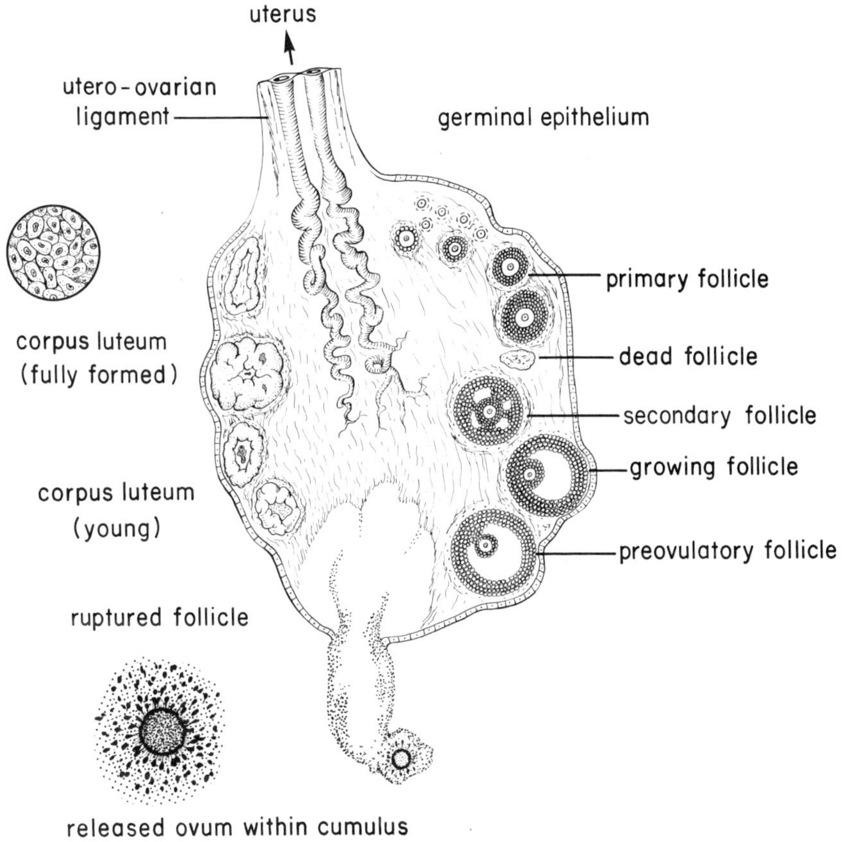

Figure 8. Sequence of events occurring in the ovarian follicle, ovulation and development of the corpus luteum during the menstrual cycle. Successive stages are depicted clockwise. Beneath the germinal epithelium is the fibrillar acellular tunica albuginea. The cellular cortex contains ovarian follicles in various stages of development. The two insets show cellular structure during ovulation and corpus luteum development.

Corpus Luteum

After ovulation, the corpus luteum develops under the influence of the luteotrophic hormone of the anterior pituitary, and secretes progesterone and estrogen. The secretion of progesterone causes the endometrium to be transformed into a luteal phase (secretory or progestational stage).

Several methods are used to detect the presence of a corpus luteum based on its increased secretion of progesterone, *e.g.*, a rise in the basal body temperature, dating of the postovulatory endometrial biopsy, a loss of the ferning pattern of dried cervical mucus and serial examination of vaginal cytology.

OVIDUCT (Fallopian tube or Uterine tube)

Structure

The oviducts are suspended in the mesosalpinx, a peritoneal fold derived from the lateral layer of the broad ligament. The oviduct can be subdivided into (1) infundibulum with its fimbriae, (2) ampulla, and (3) isthmus (Figure 5).

The infundibulum is a funnel-like structure surrounded by finger-like fimbriae arranged in a form of a fringe. The ampulla, comprising about half the oviductal length, merges with the constricted isthmus at an ampullary-isthmic junction. The isthmus is connected directly to the uterus.

Mucosa

The mucosa (epithelial lining) comprises large primary and smaller secondary folds which are particularly complex in the ampulla (Figure 6). The mucosa consists of one layer of columnar epithelial cells of three types: ciliated cells, secretory nonciliated cells, and "peg" cells, which are rod-like with very little cytoplasm.

Ciliated cells have a slender motile cilia (kinocilia) which extend into the lumen. The percentage of ciliated cells decreases gradually in the ampulla toward the isthmus. In the isthmus there are a few ciliated cells, located deep between the tall secretory cells. Secretory cells contain secretory granules, the size and number of which vary widely throughout the menstrual cycle.

Musculature

The musculature consists of an inner circular layer and an outer longitudinal layer of smooth (unstriated) muscle.

Functions

The oviduct has the unique function of conveying the eggs and sperm in opposite directions, almost simultaneously. The structure of the oviduct is well adapted to its multiple function. The fimbriae are most efficient at engulfing the eggs released from the ovary at the time of ovulation. The eggs are transported to the infundibulum through the mucosal folds of the ampulla to the ampullary-isthmic junction where fertilization and early cleavage of fertilized eggs take place. The embryos are maintained in the oviduct for three days before they are transported to the uterus. Oviductal contractions are independent of those that occur in the uterus. Contractions occur in specific patterns which propel the egg at a definite rate. The musculature undergoes various types of complex contractions: (1) localized peristalsis-like contractions originating in an isolated segment and traveling only a short distance; (2) segmental contractions; and (3) worm-like writhing of the entire oviduct. Contractions in an abovarian direction are more common than those in an adovarian direction.

Oviductal Secretions

Oviductal secretions provide a suitable environment for fertilization, cleavage of fertilized eggs and capacitation of sperm. These fluids are actively secreted from the epithelial lining of the oviduct, the secretion rate being regulated by ovarian hormones.

UTERUS

The uterus, a thick-walled, hollow, pear-shaped organ with anterposterior flattening, lies in about the center of the pelvis, anterior to the rectum and superior and posterior to the bladder. Both sides of the uterus are attached to the pelvis and abdominal walls by the broad ligament. The mulliparous uterus is about 9 cm long, 6 cm wide and 4 cm thick. The uterus is freely mobile, altering its position when the bladder or rectum become distended, and is capable of growth during pregnancy when it extends to the abdomen. This mobile position of the uterus is maintained by the broad ligaments, round ligaments, cardinal ligaments and uterosacral ligaments (Figure 5). The uterus changes size and shape during the reproductive life. In the child, the cervix exceeds the uterine body in length, but at puberty there is a growth of the uterine muscle until the uterus attains adult size. After childbirth,- the uterus is always larger than during nonpregnancy. At and after menopause, the uterus atrophies to varying degrees.

Structure

Four anatomical regions are noted along the length of the uterus from top to bottom: fundus, corpus, isthmus and cervix (Figure 9). The fundus caps the uterus. The uterine lumen exhibits maximal anteroposterior flattening, and tapers to the isthmus at the junction of corpus and cervix; then it becomes fusiform in the cervix.

The uterus is made of three major concentric layers: (1) a thin outer perimetrium or peritoneal covering; (2) a massive myometrium of smooth muscle, and connective tissue; and (3) an innermost endometrium. From the physiological standpoint, only two layers are recognized–the endometrium and the myometrium.

Endometrium

The endometrium (uterine mucosa) is composed of epithelium and stroma. A single layer of tall columnar cells lines the uterine lumen. The epithelium forms simple or branched tubular "endometrial glands" extending into the stroma, and occasionally through the myometrium. The endometrial glands, scattered throughout the endometrium, are coiled considerably, especially toward the ends. The endometrium varies in thickness from 1 to 6 mm, depending on the stage of the menstrual cycle. It also varies in vascularity with changes in ovarian hormones and pregnancy. The stroma of the endometrium contains various spiral arterioles, which are terminal vessels surrounding the glands. Progesterone causes the stromal cells to become large, pale and filled with lipid and glycogen. When fully developed, they are called decidual cells, because they are shed during menstrual bleeding or with the placenta at parturition. Bits of endometrium may implant or metastasize to various parts of the pelvis, leading to endometriosis.

Microscopic examination of an endometrial biopsy during the luteal phase can be used to access the adequacy of corpus luteal function, particularly in women with infertility problems. Aging causes a reduction in the thickness of the endometrium and this causes lowered reproductive ability in premenopausal women.

Cyclical Changes

The maturation and growth of the endometrium are under the control of ovarian estrogens and progesterone, secreted cyclically and with optimal quantitative ratios. The normal endometrium responds to the fluctuating levels of ovarian steroids, thus reflecting the stages of the ovarian cycle. This cyclic response does not involve the full thickness of the endometrium.

During menstrual bleeding, the superficial layers of the endometrium are shed. During the proliferative phase of the cycle, the glands, under the influence of progesterone, become highly coiled. During the secretory phase of the cycle, the endometrial glands become filled with the endometrial fluid which flows through the openings of the uterine glands. During pregnancy, the progestational endometrium increases in thickness to allow the implantation of the blastocyst *in utero.*

Myometrium

The myometrium, the muscular portion of the uterine wall, consists of two layers of smooth muscle: a thin outer longitudinal layer continuous with that of the oviduct and vagina, and an inner layer arranged in oblique spirals. Between them lies a vascular layer made of blood and lymph vessels, nerves and connective tissue. The myometrium is arranged in intricate layers of interlacing bundles of smooth muscle cells which respond to the action of oxytocin, a hormone secreted by the posterior lobe of the pituitary gland. The frequency and patterns of myometrial contraction vary with different stages of the menstrual cycle. The motility of the myometrium is of major physiological

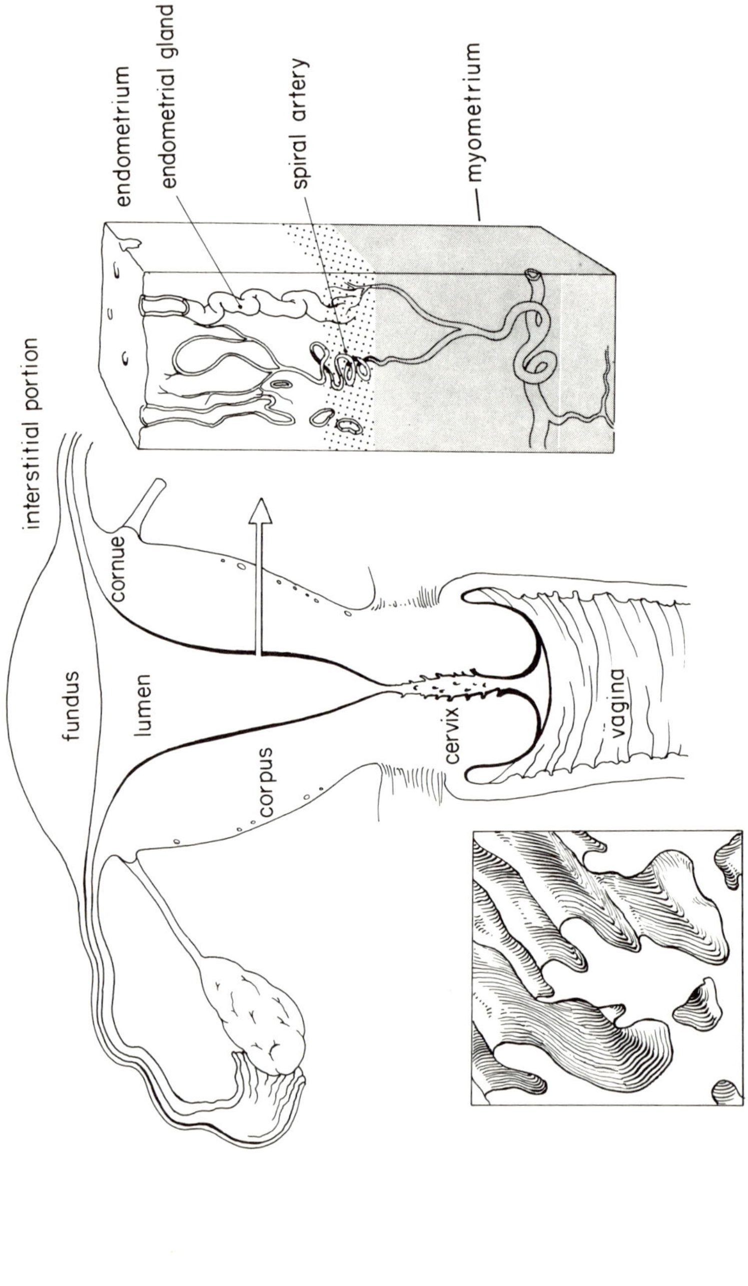
endometrium
endometrial gland
spiral artery
myometrium
interstitial portion
cornue
fundus
lumen
corpus
cervix
vagina
arbor vitae

Figure 9. Diagrammatic illustrations of various segments of the uterus (fundus, cornue, corpus), and uterine cervix showing details of cervical crypt (arbor vitae). The endometrium is made of three layers.

(a) Zona compacta: surface layer surrounding openings of endometrial glands, during luteal phase forms one-third of total thickness.

(b) Zona spongiosa: loose stroma with endometrial glands.

(c) Zona basalis: relatively thin.

The basal layer of the endometrium, which is less sensitive to hormonal changes than superficial layers, remains unchanged throughout the menstrual cycle. Blind ends of endometrial glands regenerate after menstruation or childbirth. This layer cannot be removed with a curette. The myometrium (shaded area) is primarily made up of smooth musculature and connective tissue.

significance for the rapid transport of sperm from the cervical canal to the oviduct, for the expansion of the uterus during pregnancy, and for the expulsion of the fetus during parturition.

Function

The uterus and endometrial fluid play a major role in reproduction. As sperm are transported through the uterine lumen to the oviducts, they undergo capacitation in the endometrial secretions. Before implantation, the developing blastocyst is nourished by the uterine secretions. After implantation, the developing embryo and placenta depend on an adequate vascular supply within the uterus. Throughout gestation, the physiological properties of the endometrium and its blood supply are important for the survival and development of the fetus.

Sperm Transport

At coitus, the contractions of the myometrium are essential for the transport of sperm from the site of ejaculation to the site of fertilization. Large numbers of spermatozoa aggregate in the endometrial glands. The physiological and/or immunological significance of this phenomenon is not known.

Large clusters of white blood cells (leukocytes) appear in the lumen of the uterus in response to the presence of either a foreign body such as an IUD or sperm, after intercourse. Such leukocytes play a role in phagocytosis (engulfing with subsequent destruction) of unnecessary sperm and perhaps capacitation of sperm.

Size Changes

The uterus is capable of undergoing tremendous changes in size, structure and position to accommodate the needs of the growing conceptus (fetus and placenta) during pregnancy. Its great contractile power remains dormant until the time of parturition, when contraction is essential for fetal expulsion. Following parturition, the uterus almost completely regains its former size and condition through a process called postpartum involution.

CERVIX

Structure

The cervix, a sphincter-like structure which projects caudally into the vagina, is characterized by a thick wall and constricted lumen. The cervix is made of fibrous and elastic tissues; the upper part contains smooth muscles condensed in a circular pattern to form the internal os or sphincter of the uterus. The cervix does not contain any glands as has been assumed erroneously. The cervical mucosa is made of primary and secondary mucosal crypts which run in transverse, longitudinal and oblique directions (Figure 9). This pattern provides an extensive secretory and storage surface. The cervical lining is made of mucus-secreting columnar epithelium (Figure 7). Scattered ciliated cells are found individually or in clusters among the mucus-secreting cells. These secretory cells are filled with abundant secretory granules, whereas the ciliated cells contain only sparse granules. The cervical epithelium is supported by a fibrous groundwork.

Histological changes in the endocervical epithelium that occur cyclically are subtle. Columnar cells reach their maximal height at midcycle, just prior to ovulation.

Function

The cervix prevents intruders from entering the endometrial lumen. It is closed tightly except during midcycle, when the external os is dilated and filled with watery cervical mucus, which facilitates sperm transport to the uterus.

The cervix secretes the cervical mucus which, due to its unique rheological properties, plays a major role in sperm transport. The cervical crypts, lined by mucus-secreting cells, are responsible for the formation of micellar rods. The cervical crypts are considered as sperm reservoirs since motile sperm are stored there for as long as one week following intercourse. Sperm are then released and migrate toward the uterine lumen.

Cervical mucus has a bactericidal effect which becomes reduced at the time of ovulation, probably as a result of increased volume of mucus which causes a dilution of lysozymal or other bacterial components. During ovulation, bacteria not only remain alive, but also are able to migrate through the mucus.

Cervical Mucus

Cervical mucus consists of rod-shaped macromolecules of mucin, synthesized by the secretory cells lining the cervical crypts and cervical canal (Figure 10). The mucus that accumulates in the vaginal pool may contain endometrial, oviductal, follicular and peritoneal fluids as well as leukocytes and cellular debris from uterine, cervical and vaginal epithelia. The secretion of cervical mucus is stimulated by estrogens and inhibited by progesterone. Cyclical qualitative changes in the cervical mucus throughout the menstrual cycle, and cyclical variations in the arrangement and viscosity of these macromolecules cause periodic changes in the penetrability of sperm in the cervical canal. Optimal changes of cervical mucus properties, such as an increase in quantity, viscosity, ferning pattern (crystallization) of dried mucus and pH, and decrease in viscosity and cell content, occur during midcycle, and are reversed during the luteal phase when sperm penetration in the cervix is inhibited.

Under the influence of estrogens, the macromolecules of mucus are oriented so that the spaces become larger than under the influence of progestins. During pregnancy, a highly viscid, nonferning, thick and turbid mucus occludes the cervical canal, acting as a barrier against sperm transport and uterine infections. Before parturition, the cervical plug of pregnancy liquefies, and the cervix dilates to permit the expulsion of the fetus and placenta.

Biological Characteristics

The biochemical, biophysical and physiological characteristics of cervical mucus are summarized in Table III. Cervical mucus is made up of high- and low-viscosity components. High-viscosity glycoprotein macromolecules, arranged in long chains, join in linear filaments which aggregate into networks or micelles (Figure 10). These micelles contain the water and solutes that comprise the low-viscosity component. The network of macromolecules

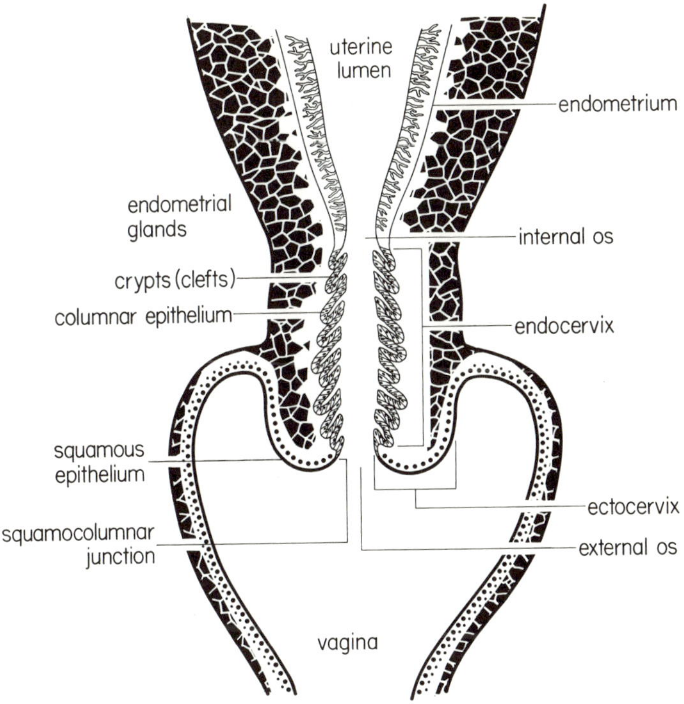

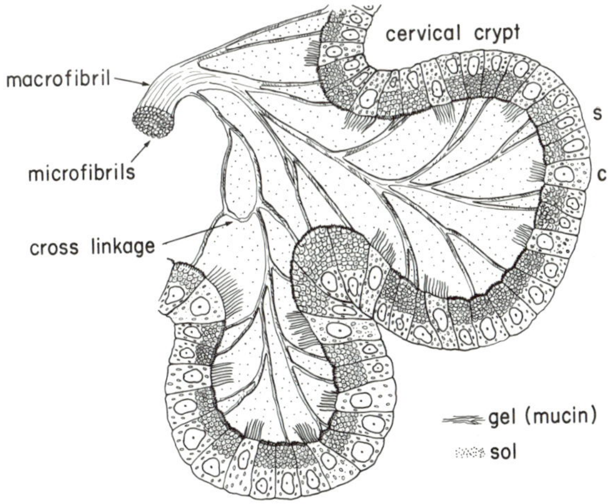

Figure 10. (a) Anatomy of the cervix. (b) Diagrammatic illustration of formation of microfibrils and macrofibrils of mucin from the secretory cells within cervical crypts. The beat of cilia may facilitate the release of mucin from the adjacent secretory cells. The mucin makes the framework of the "gel" whereas the "sol" allows free transport of spermatozoa.

Table III

Biochemical, Biophysical and Physiological Characteristics of Cervical Mucus

Characteristics	*Parameters*	
Biochemical components	Water content	95-98% during midcycle 85-90% during rest of cycle
	Dry matter (glycoprotein)	25% amino acids arranged in long polypeptide chains 75% carbohydrates: galactose, glucosamine, fructose and sialic acid
	Plasma proteins	Serum albumin, globulin, lipoprotein, immunoglobulin
	Enzymes	Phosphatase, amylase, phosphorylase, glucouronidase
	Inorganic salts	NaCl, Mg, K, P
Rheological characteristics	Mucusis hydrogel made of two components: 1. High-viscosity component (gel phase) Glycoprotein network Network of glycoprotein macromolecules arranged in long chains 2. Low-viscosity component (sol phase) Electrolytes, organic compounds of low molecular weight, *e.g.*, glucose, amino acids and soluble proteins in plasma-like fluid	
Biophysical properties	Spinnbarkeit (stretchability) Tach (stickiness) Ferning (crystallization upon air-drying) Elasticity Viscosity Thixotrophy	

Table III, continued

Characteristics		*Parameters*
Changes during midcycle under estrogen influence	Rheological changes	Increased Spinnbarkeit, decreased viscosity
	Biochemical changes	Decreased % of soluble proteins, decreased mucin-bound sialic acid
	Physiological changes	Increased sperm penetration, increased production, increased water content (98-99%), decreased leukocyte infiltration
Daily secretion rate	600 mg/day during midcycle 20-50 mg/day during rest of cycle	

is responsible for the rheological characteristics of the mucus. The fluid properties of mucus relate to the amount of free water molecules between the micelles.

Due to its biophysical properties, cervical mucus has several rheological characteristics (Table III). Ferning refers to a crystallization pattern observed when cervical mucus, under the influence of estrogen, is allowed to dry and then is examined microscopically. Ferning results from the combination of NaCl and protein in a distinct proportion. Spinnbarkeit is measured as the length of a thread of mucus which can be drawn between a microscope slide and a cover slip. Viscous mucus is converted within a few days to clear, watery mucus with low viscosity and excellent Spinnbarkeit by the daily administration of estrogens. Progestational contraceptives can rapidly reverse this change.

Cyclical Changes

During midcycle, the micelles of cervical mucus permit sperm transport toward the uterine cavity (Figure 10). During the luteal phase, however, the micelles form a dense network with very narrow channels, which inhibit sperm migration. Under estrogen stimulation during midcycle, cervical mucus undergoes dramatic rheological, biochemical and physiological changes (Table III).

Cyclical changes in rheological properties of cervical mucus, associated with hormonal milieu, have been used in attempts to predetect ovulation clinically. Reliability of these parameters is only high when the mucus is examined serially, whereas reliability is lost in the presence of cervical infection.

VAGINA

Structure

The vagina is a fibromuscular canal located between the bladder and the rectum, and extending from the vestibule to the cervix, around which it is attached, to form the fornices (Figure 11). The vagina is 7 to 10 cm long in its anterior wall and 12 to 15 cm long in its posterior wall. Normally, the anterior and posterior walls are in contact so the lumen is represented by an H-shaped slit. The vagina is capable of great distention as during intercourse or childbirth.

The vaginal surface is corrugated by the development of numerous transverse rugae (folds) which are particularly prominent on the anterior and posterior walls in the lower two-thirds of the vagina. The rugae have physiological significance during intercourse and delivery. They appear during puberty in response to estrogen stimulation and disappear in postmenopausal women.

The vaginal surface, made of numerous rows of stratified squamous epithelium, consists of three layers: a superficial cornified layer (Figure 10), a functional layer and a basal layer. On the surface the cells are large, flat and with pyknotic nuclei, whereas at the base, the cells are small and round, with large nuclei. The vaginal surface is continuously shedding old cells and replacing them with new ones. The vaginal surface undergoes cyclical changes during pregnancy. The relative numbers of cell types in a vaginal smear, as observed microscopically, are used as a criterion of estrogen status of the woman. In postmenopausal

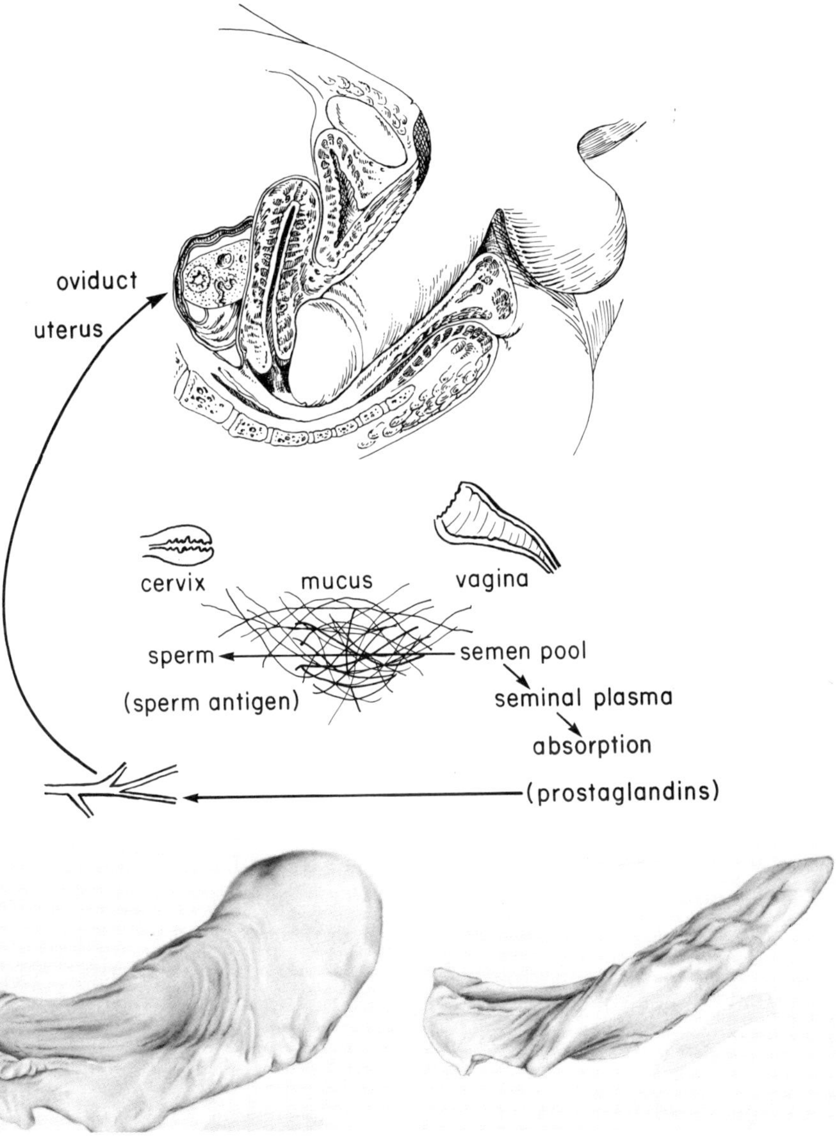

Figure 11. Top: In the lithotomy position, the posterior vaginal fornix becomes the lowest part of the vagina. This anatomical relationship favors sperm transport from the vaginal pool to the cervix. Note the external os of cervix resting on posterior fornix, which accommodates the erected penis.
Bottom: Vaginal mold. Note distention of posterior fornix to allow for semen deposition during intercourse. [From Hafez, E. S. E. *Am. J. Obstet. Gynaecol.* 125:573 (1978).]

women not taking estrogen supplements, the vaginal surface becomes thin, smooth and atrophic. Estrogen therapy causes progressive shift to the superficial cells.

The muscular coat of the vagina consists of a thick, inner circular layer and a thin, outer longitudinal layer, the latter continuing for some distance into the uterus. This musculature is supplied with abundant blood vessels and nerve bundles. The vagina is less sensitive than the clitoris and vulva. The outer third of the vagina contains nerve endings responsive to sexual stimulation whereas the inner third of the vagina has very few nerve endings. Near the opening, there is a muscular sphincter which contracts rhythmically during orgasm. The muscular wall of the vagina may be stretched or torn during childbirth, and becomes atrophied and less vascular as estrogen levels decrease, so that such herniations as rectocele and cystocele are common in multiparous, postmenopausal women.

Squamocolumnar Junction

The squamocolumnar junction refers to the area of transformation from the stratified epithelium of the vagina to columnar epithelium of the cervix. There seems to be functional interaction between stratified squamous and columnar epithelial cells, since this junction is the initial site of squamous cell carcinoma. Age-related changes in the distribution of cervical epithelium cause varying degrees of eversion of the mucosa. At puberty, the squamocolumnar junction may extend beyond the external cervical os, so that columnar epithelium cover the vaginal aspect of the ectocervix. In postmenopausal women, this junction withdraws to within the endocervical canal.

Bartholin's Glands

The glands of Bartholin (greater vestibular glands), the size of a pea, are composed of racemose glands which resemble salivary glands. The gland is embedded in the posterior part of the bulb of the vestibule, a mass of erectile tissue lying in the labium minus. The duct, about 2 cm long, opens into the vestibule at the junction of the labium minus and the hymen. During sexual excitement, the glands produce a colorless mucin-like secretion.

Function

The vagina has several functions during menstruation, intercourse, and during parturition (Figure 12). The vagina is an excretory canal for menstrual discharges, and provides an opportunity for the clinician to examine the internal genitalia.

Intercourse

During intercourse the vagina is a dilatable organ. Semen is deposited into the vaginal fornix (arch formed by the projecting cervix) and sperm are transported through the external os of the cervix (Figure 11).

The direction of the vagina is optimal for intercourse since it extends backward and upward from the vulva to the uterus to accommodate the erect, upward-angulated penis. During coitus the larger vaginal fornix becomes the lowest part of the vagina. This anatomical situation favors sperm transport from the vaginal pool to the cervix since the external cervical os rests in the posterior fornix. The vaginal ridges reinforce the gripping effect of the levator ani and vaginal constrictor muscle during coitus. This phenomenon, accompanied by an encircling action vulvovaginal clasp, and fixation of the fully erect penis, facilitates subsequent orgasm.

Sexual excitement is associated with a sweating phenomenon: droplets of transudates appear on the vaginal rugae to provide sufficient lubrication for coital receptivity. This process seems to be regulated by an intricate system of venous plexuses around the vagina, vulva, rectum and bladder. With sexual stimulation, the accumulation of blood in the plexuses causes vaso-congestion of the vaginal walls and increased transudation.

Parturition

During parturition the fetus and placenta are expelled through the birth canal. The vaginal rugae undergo considerable distention and stretching during delivery. Vaginal rugation are much more prominent in nulliparous than multiparous women. During the first stage of labor, biochemical changes occur in the colloids of the vaginal wall making them more hydrophilic and thus capable of undergoing considerable dilation.

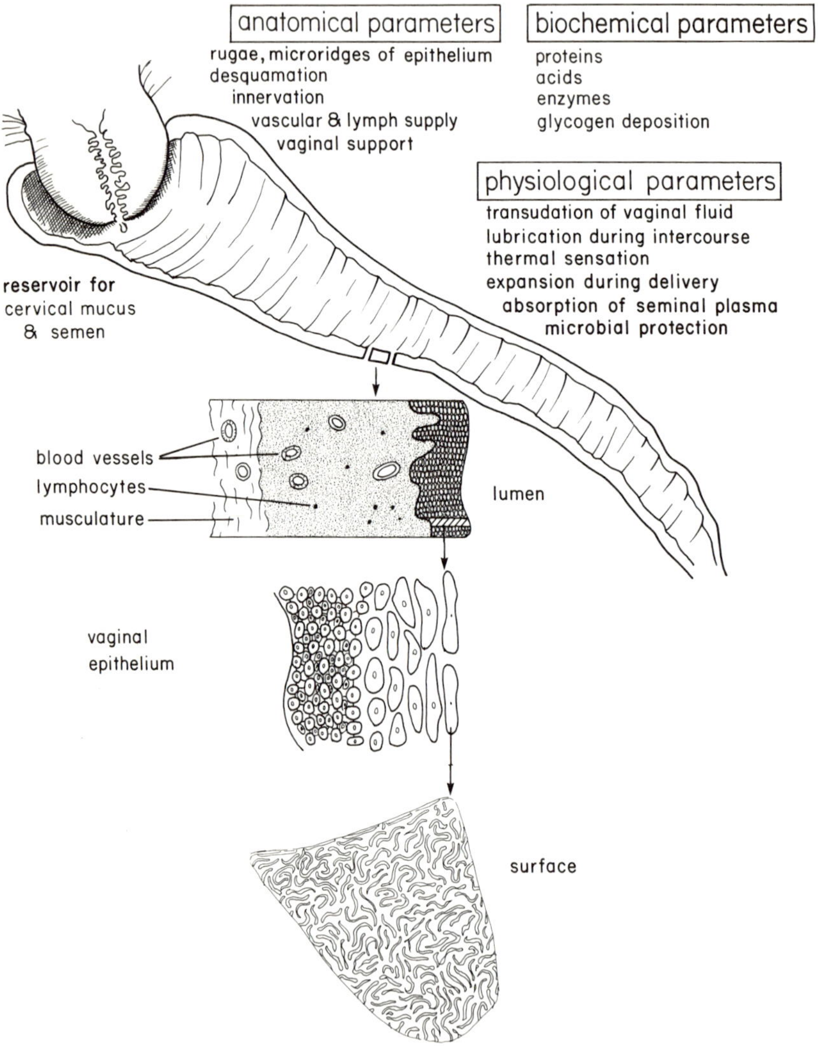

Figure 12. Anatomical, physiological and biochemical parameters influencing the role of the vagina in reproductive processes:

(a) as a copulatory organ;

(b) as semen pool for cervical reservoir; and

(c) distention during courtship, intercourse and parturition. [From Hafez, E. S. E. *Am. J. Obstet. Gynaecol.* 125:573 (1978).]

Protection

The vaginal epithelium is relatively impervious to microorganisms because of its multilayered structure and the cornification of the superficial cells. The acidic pH of the vagina provides a natural resistance to infection by preventing excessive bacterial growth. Before puberty and after the menopause, the pH is less acidic, increasing susceptibility to infection.

Vaginal Fluid

Unlike the cervix, the vagina does not secrete mucus. Semifluid, whitish and ordorless vaginal discharges are composed of cervical mucus, vaginal epithelial transudate, uterine secretions, epithelial cells, leucocytes and bacteria, primarily Döderlein's bacilli. The amount and consistency varies throughout the menstrual cycle. During midcycle, vaginal discharges are accentuated and may be profuse in some women due to excessive secretion of cervical mucus before ovulation. Vaginal discharges are acidic (pH 4 to 5) as a result of the lactic acid produced from the glycogen in the epithelial cells, under the influence of lactobacillus of Döderlein, the latter being a large gram-positive rod bacteria found in abundance in the normal vagina.

During sexual excitement, mucus-like lubricating fluid is formed from: (1) transudation of serum and vaginal tissue fluid, (2) secretions from Bartholin's glands near the vaginal opening, and (3) secretions from Skene's glands located near the opening of the urethra into the vaginal vestibule.

Vaginal glycogen undergoes rapid turnover under the influence of increasing levels of estrogens. Glycogen-rich cells exfoliated into the vaginal fluid also increase. Vaginal fluid contains a variety of enzymes, several amino acids, lactic acid, glucose, maltose and various carbohydrates. The concentration of some of these components undergo remarkable cyclical changes throughout the menstrual cycle, *e.g.*, urea, acetic acid and lactic acid.

Vaginal Odor

Various pleasant vaginal odors have been noted, *e.g.*, fruity, spicy, aldehyde and floral. Unpleasant odors include amino burnt,

burnt earth, acidic medicinal, phenolic, musty and cheesy. There are remarkable variations in vaginal odors amongst women, and at various stages of the menstrual cycle of a given woman.

Vaginal odors are derived from decomposition products of microorganisms, desquamated cells from the urogenital tract, vaginal secretions, cervical mucus and semen. Several microorganisms that are situated in the vagina may be partially responsible for vaginal odors. For example, pyogenic organisms may cause abnormal purulent vaginal discharges, *e.g.*, *Streptococcus pyogenes, Staphylococcus aureus, Escherichia coli, Trichomonas vaginalis,* or the yeast-like fungus *Candida albicans.* Various factors influence microbial growth in the vagina, *e.g.*, abrupt differences in pH, cellular glycogen content and biochemical properties of cervical mucus.

Little is known about the possible effects of pleasant vaginal odors on sexual responses of man, whereas malodors of vaginal fluids may interfere with normal sexual response.

EXTERNAL GENITALIA

Until the seventh week of pregnancy, the external development of the genitalia appears the same in the male and female fetuses. In the presence of fetal testosterone, male development occurs, whereas in the absence of these androgens, female development occurs.

The female external genitalia consist of several parts: mons veneris, labia majora, clitoris, labia minora, vestibule, glands of Bartholin and hymen. The mons veneris, a pad of fat, which lies over the symphysis pubis, is covered with the pubic hair that grows at puberty. The labia majora, two folds of skin which enclose the vaginal opening, contain fatty tissue and extend on either side from the mons veneris and merge posteriorly into the perineum. Folds of skin from the labia minora form the prepuce of the clitoris, which contains erectile tissue attached to the pubic arch. The labia minora contains fibrous tissue and several blood capillaries constituting erectile tissue. The skin of the labia, containing sebaceous glands but no hair follicles, represents the transition from true skin to the squamous epithelium lining the vestibule and vagina.

The external genitalia are target organs for estrogens. As the result of decreasing estrogen levels after menopause, the external genitalia become atrophic and predisposed to several lesions. The elasticity of the labia minora and the clitoris decrease and the glands undergo atrophy. This atrophy, however, does not necessarily interfere with intercourse.

Glands of the Vestibule

Several glands located in the subcutaneous and connective tissue of the external genitalia secrete fluids into the vestibule. During sexual excitement, Bartholin's glands secrete at the base of the vestibule a whitish-gray fluid. Small alveolar vestibular glands, distributed throughout the walls of the vestibule, secrete mucus. The largest of these glands open through two to four ducts between the vaginal introitus and the urethra. The external os of the urethra is located in the anterior segment of the vestibule. The secretions of the sebaceous glands in this part protect the vulnerable tissues against urine. Skene's (paraurethral) glands have ducts which open laterally to the urethra and may harbor microorganisms.

Clitoris

The clitoris is the embryological homologue of the penis and consists of erectile cavernosus bodies of a few millimeters in length and ending in the rudimentary glans clitoris. The glans can be viewed externally between the folds which are formed as a result of the dividing branches of the labia minora. The clitoris, being generously endowed with nerve endings, is a delicately sensitive organ which channels sexual excitement. Upon sexual stimulation, the clitoris is engorged with blood, undergoes partial "erection" and may enlarge to 2.5 cm in length.

Hymen

The normal hymen is a circular or cresentric fold of squamous epithelium in the form of a thin, weak and elastic membrane partially covering the opening of the vagina in virgin females. The opening allows passage of menstrual blood. The abnormal hymen may be perforate, imperforate, sieve-like, labiform or bipartus. The hymen is stretched or ruptured at the first intercourse, leaving carbuncle-like remnants which retain semen in the vaginal pool, thus increasing the chances of conception. Some girls are born without hymens, while in others the hymen may be stretched or ruptured by causes other than intercourse, *e.g.*, athletics or insertion of sanitary tampons. The fingers can be used to stretch the hymen prior to first intercourse.

CHAPTER 3

HORMONES OF REPRODUCTION

Human reproduction is regulated by hormones from three separate structures, each with its distinct secretions: (1) the hypothalamus, which secretes hypothalamic-releasing and -inhibiting hormones; (2) the pituitary, which secretes gonadotropic hormones (gonadotropins); and (3) the gonads (ovary and testis), which secrete steroid hormones. The hypothalamus receives input from exteroceptive stimuli to regulate the internal hormonal state (Figure 13). Hypothalamic hormones are transported via the portal system to the pituitary gland, which in turn secretes appropriate tropic (stimulating) hormones. The gonads, stimulated by gonadotropins, release steroid hormones to regulate the structure and function of the reproductive tract (Figure 14). Gonadal hormones also exert an effect on the pituitary gland and/or the hypothalamus as part of a feedback system. The sequence of these hormonal interactions are well synchronized and coordinated to ensure a complete reproductive cycle.

Primary hormones that regulate and control reproduction include the hypothalamic hormones; follicle-stimulating hormone (FSH), luteinizing hormone (LH) and prolactin secreted by the anterior lobe of the pituitary gland; oxytocin secreted by the posterior lobe of the pituitary; androgens secreted by the testes; estrogens and progesterone secreted by the ovary; and placental hormones secreted during pregnancy (Table IV). Other secondary hormones indirectly related to reproduction are secreted by the pituitary, thyroid, parathyroid, adrenal cortex and pancreas (Table V).

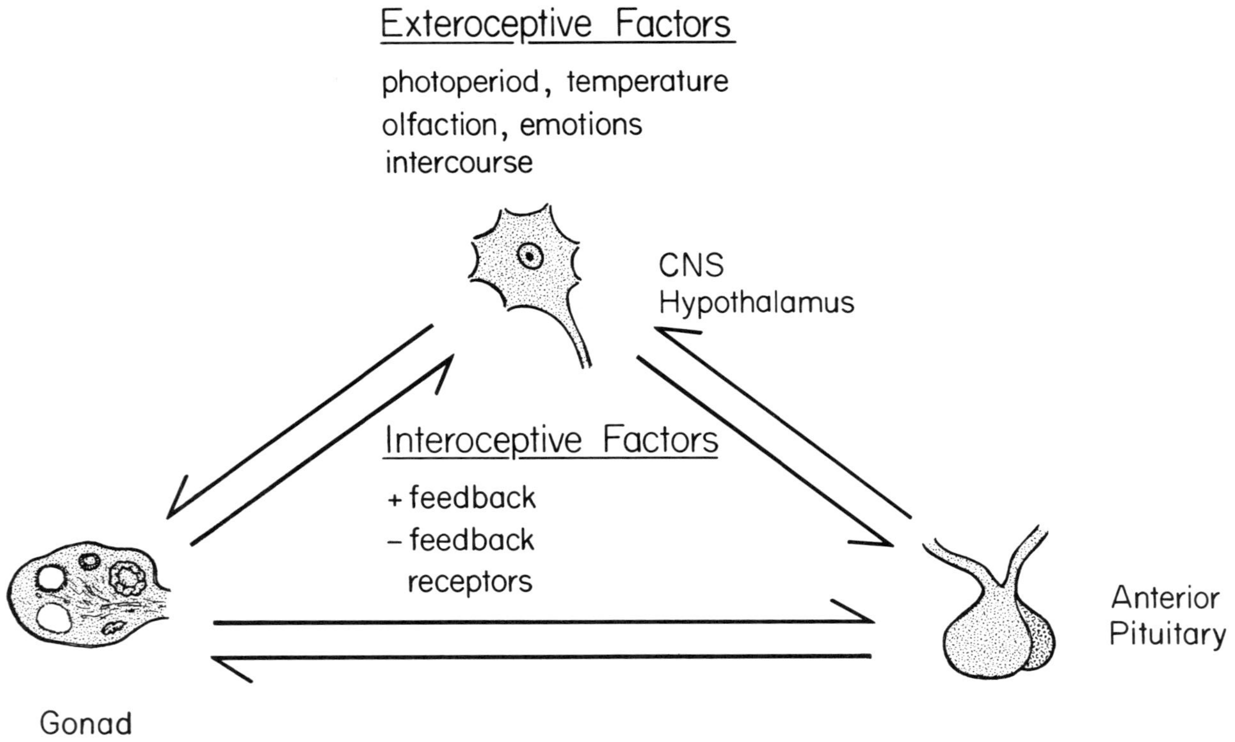

Figure 13. The endocrine relationships of exteroceptive and interoceptive factors as they affect the functions of the hypothalamus (part of the brain), pituitary gland and gonads.

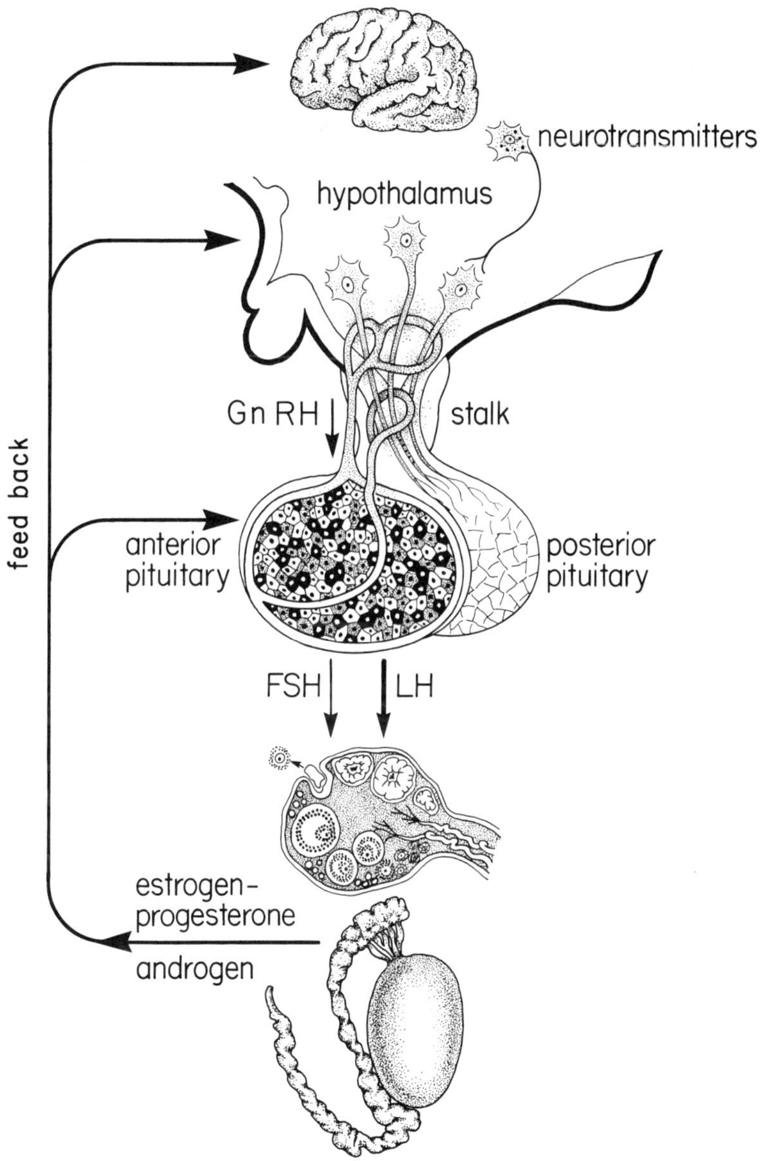

Figure 14. Schematic presentation of the endocrine and neuroendocrine relationship among hypothalamus, pituitary gland and gonad (ovary-testis). Hypothalamic neurosecretory materials (GnRH) are transported by the portal blood capillaries to the cells of the anterior pituitary. FSH and LH stimulate the gonads. Estrogens and androgens secreted by gonads exert a feedback on the pituitary and hypothalamus.

Table IV

Primary Hormones which Regulate and Control Reproduction

Source	*Hormone*	*Some Functions*
Hypothalamus	ACTH-RH (adrenocorticotropin-releasing hormone)	Release of ACTH
	Gn-RH & LH-RH (gonadotropin-releasing hormone)	Release of LH and FSH
	GRH (growth hormone-releasing hormone)	Release of growth hormone
	TRH (thyrotropin-releasing hormone)	Release of TSH and prolactin
	PIF (prolactin inhibiting factor)	Inhibits prolactin release
Anterior Pituitary	FSH (follicle-stimulating hormone)	Spermatogenesis In synergism with androgens, stimulates Sertoli cells Growth and maturation of ovarian follicle in ovary
	LH (ICSH) (luteinizing hormone)	Stimulates testicular Leydig cells to secrete androgens Ovulation Stimulates theca interna cells of the ovary to luteinize
	LTH-prolactin (luteotropic hormone)	Progesterone release Lactation Stimulates testosterone secretion
Posterior Pituitary	Oxytocin	Parturition; uterine contractions; milk let-down; sperm transport in female reproductive tract

Table IV, continued

Source	*Hormone*	*Some Functions*
Testis	Androgens	Maintain male reproductive duct system and secondary sexual characteristics Descent of testes, development of male sexual behavior Stimulate spermatogenesis, anabolic effect, promote spermatogenesis in synergism with FSH Feedback control of gonadotropins
Ovary	Estrogens	Stimulate growth of oviduct, uterus and vagina Proliferation of endometrium Stimulate female secondary sexual characteristics Stimulate growth of ducts in mammary glands Feedback control of gonadotropins
	Progesterone	Transform proliferating endometrium to secretory endometrium Induce ovulation in synergism with estrogens Inhibit ovulation Initiate implantation of blastocysts Maintain pregnancy in synergism with estrogen

Table V

Secondary Hormones of Reproduction

Source	*Hormone*	*Some Functions*
Placenta	hCG (human chorionic gonadotropin)	LH-like
	Estradiol	see Ovary
	Progesterone	see Ovary
Anterior Pituitary	STH (somatotropin hormone)	Body growth; protein synthesis
	TSH (thyroid-stimulating hormone)	Stimulates thyroid gland Thyroxine release and iodine uptake by thyroid
	ACTH (adrenocorticotropic hormone)	Stimulates adrenal cortex Release of adrenal corticoids
Posterior Pituitary	ADH (vasopressin) (antidiuretic hormone)	Water balance
Thyroid	Thyroxine	Body growth; development and maturation; oxidation of feeds
Parathyroid	Parathormone	Calcium and phosphorus metabolism
Adrenal Cortex	Aldosterone	Electrolyte and water metabolism
	17-OH corticoids (cortisone) (cortisol)	Carbohydrate, protein and fat metabolism
Pancreas	Insulin	Carbohydrate, fat and protein metabolism

HYPOTHALAMUS

The hypothalamus, which occupies only a very small portion of the brain, consists of the region of the third ventricle, extending from the optic chiasma to the mammillary bodies. The hypothalamus, an important neuroendocrine center, releases several hormones (Table IV). Naturally occurring releasing hormones, extracted from thousands of ovine and porcine hypothalami have been used to determine the biochemical structure of these hormones. For example, the hypothalamic gonadotropin-releasing hormone (GnRH) is a decapeptide composed of 10 amino acids. Several of the hypothalamic hormones have been synthesized and marketed for clinical use.

Function

The hypothalamus regulates various vital automatic processes, *e.g.*, appetite, heart rate, temperature control, sexual behavior and neuroendocrine activity. Various centers in the hypothalamus integrate physiological signals in the body, such as central nervous system, metabolic status, functional activity of target glands, and the internal environment. The hypothalamus then responds by elaboration of specific releasing hormones.

Thus, the hypothalamus acts as a processing and integrating center for received information and translates it into a neurohormonal signal which evokes physiological responses.

The main hormones released by the hypothalamus that regulate reproduction include the gonadotropin-releasing hormone (GnRH or LH-RH), adrenocorticotropin-releasing hormone, (GRH) and prolactin inhibiting factor (PIF) (Table IV). The hypothalamus is also the source of oxytocin and vasopressin, the hormones stored in the posterior lobe of the pituitary gland.

Mode of Action

Specific hypothalamic neurons synthesize and elaborate peptide hormones, which control the release and synthesis of hormones of the anterior lobe of the pituitary gland.

Although the anterior pituitary is under hypothalamic control, such regulation is not achieved by means of neural connections, but rather by the transport of hypothalamic neurohormones from the hypothalamus to the anterior pituitary via a vascular connection, the "hypothalamo-pyprophysial portal system." Interruption of these blood vessels in the stalk of the pituitary gland leads to gonadal atrophy. If, however, the stalk of the pituitary is allowed to revascularize, gonadal functions can resume. On the other hand, electrical or hormonal stimulation of specific nuclei of the hypothalamus can regulate gonadal function.

Clinical Uses of Hypothalamic Releasing Hormones

GnRH has been used for the treatment of delayed puberty and various infertility problems in both men and women. GnRH and TRH injected intravenously in microgram quantities provide a valuable diagnostic tool to distinguish between hypothalamic and pituitary defects. TRH is used to increase milk production by nursing mothers. The therapeutic efficacy of somatostatin in acromegaly and diabetes mellitus is under investigation.

PITUITARY GLAND (HYPOPHYSIS)

The pituitary gland is located in the sella turcica, a bony depression at the base of the brain. The gland is subdivided into two distinct anatomical parts: an anterior lobe and posterior lobe. The anterior pituitary contains three cell types which vary in proportion according to sex, age, and reproductive stage and pregnancy. Each cell type secretes one or two hormones, *e.g.*, luteinizing hormone (LH), follicle-stimulating hormone (FSH) and prolactin.

FSH and LH

Human LH and FSH are glycoproteins with molecular weights of approximately 28,000 and 35,000, respectively. Each glycoprotein hormone consists of two dissimilar subunits, designated alpha (α) and beta (β), which are united by several weak forces (Figure 15). The alpha subunit seems to be similar for

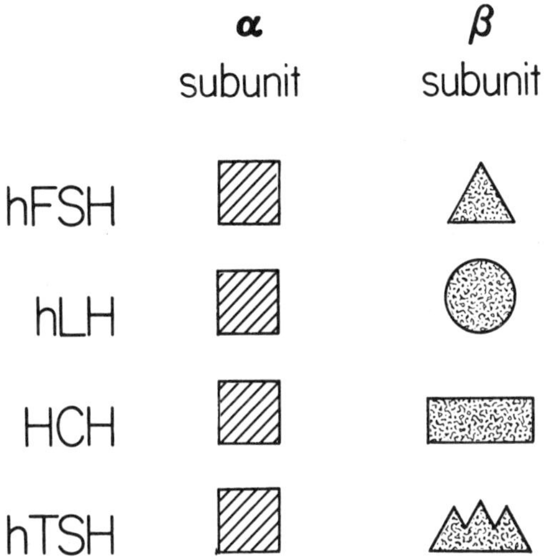

Figure 15. Subunits of pituitary protein hormones. Note that the biological activity is determined by β subunits whereas α subunits are similar in structure.

the various glycoprotein hormones of a given species, whereas the beta subunits appear to be hormone-specific. Either subunit alone has little biological activity, but the activity is restored when the two subunits are united.

The actions of FSH and LH are confined to the gonad. In men, LH stimulates the proliferation and function of the interstitial cells of the testes to secrete testosterone whereas FSH acts on Sertoli cells and/or primary spermatogonia during spermatogenesis (sperm production). In women, FSH promotes ovarian follicular growth in the ovary. LH in combination with FSH has multiple functions. It is responsible for stimulating estrogen secretion and, in higher concentration, for ovulation itself. It is also responsible for luteinization of the ruptured follicle.

Due to the cyclic function of the female gonad, the release of pituitary hormones in blood serum is cyclical in nature in women but not in men. FSH and LH are not released in the blood at a constant rate, but rather by a sequence of small secretory bursts which result in abrupt increases in concentration at some hourly intervals.

Human Chorionic Gonadotropin (HCG)

Chorionic gonadotropin secreted by the placenta is present in appreciable amounts in the blood of pregnant women. Although it is secreted throughout pregnancy, maximal secretion of HCG, reached at one month gestation, is followed by a rapid decline. HCG seems to maintain the corpus luteum, which secretes progesterone.

Diagnosis of pregnancy is based on the detection of HCG, which appears in the urine about the time of implantation. Certain malignancies, such as chorioepithelioma and hydatiform moles in women are associated with high levels of urinary gonadotropins.

Gonadotropins and Ovulation Induction

Various preparations of gonadotropic hormones (Table VI) are used to induce ovulation and pregnancy in anovulatory women. Excessive dosage of FSH, a higher concentration of LH than FSH in the preparation, or prolonged treatment with HCG, may cause multiple ovulation and multiple pregnancy. To avoid multiple ovulation, the patients are carefully selected, and their endocrine profile monitored throughout the treatment.

Prolactin (Luteotrophin, LTH)

Some authors provided evidence of the presence of a purely prolactin hormone in man, whereas others believe that prolactin is merely the exclusive property of human growth hormone (HGH). The exact role of prolactin in regulation of the function of the human gonads is not clear.

Growth Hormone (hGH)

Growth hormone is essential for body growth, and influences protein, fat and carbohydrate metabolism. It has no specific

Table VI

Gonadotropic Hormones Clinically Used to Stimulate the Ovaries and to Induce Ovulation and Corpus Luteum Formation in Anovulatory Women

	Preparation	*Source of Extraction*
HCG	Human chorionic gonadotropin	Urine of pregnant women
HPG	Human pituitary gonadotropin	
hFS (FSH)	Human follicle–stimulating hormone	Postmortem human pituitary gland
hLH (LH)	Human luteinizing hormone	
HMG	Human menopausal gonadotropin	Urine of postmenopausal women
PMSG	Pregnant mare serum gonadotropin	Serum of pregnant mares
Various preparations	Animal pituitary gonadotropin	Pituitary glands of animals

target glands but acts on several tissues in the body. GH secretion is controlled mainly, but not exclusively, by the hypothalamus.

STEROID HORMONES

According to their biochemical structure and function, the steroids are divided into three major groups: androgens, estrogens and progestins. All steroid hormones have a four-carbon ring skeleton known as cyclo-penteno-phenanthrene derivatives. The basic structural formula for natural estrogen contains 18 carbon atoms (Figure 16). Estradiol is the primary estrogen, with estrone and estriol representing other metabolically active estrogens. Androgens are 19-carbon steroids with a hydroxyl or oxygen at positions "3" and "17" and a double-bond at position "4." The androgens are called 17-ketosteroids when oxygen is found at position "17."

Several enzymes regulate the biosynthesis of steroid hormones from cholesterol in several endocrine glands, *e.g.*, the ovary, testis, adrenal glands and placenta. The testis primarily synthesizes androgens, whereas the ovaries synthesize two major types of steroids: 18-carbon estrogens and the 21-carbon progestins. Cholesterol in steroidogenic cells may be presynthesized *de novo* or derived from pools within the plasma, which are largely contributed by the liver. In the blood plasma, a large portion of each steroid hormone is bound to albumin, a plasma protein with low affinity and high capacity for steroids. Another portion of the steroid hormone is bound to one or more specific proteins with high affinity. The half-life of naturally occurring steroids in the body is very short. Therefore, several steroids with modified biochemical structure have been synthesized for clinical use.

Steroid hormones cause a variety of physiological responses in target tissues, *e.g.*, cell division, tissue differentiation, growth, synthesis of specific proteins and contraction of smooth muscle. These responses are responsible for various reproductive processes, *e.g.*, sexual behavior, sexual receptivity, priming the uterus for

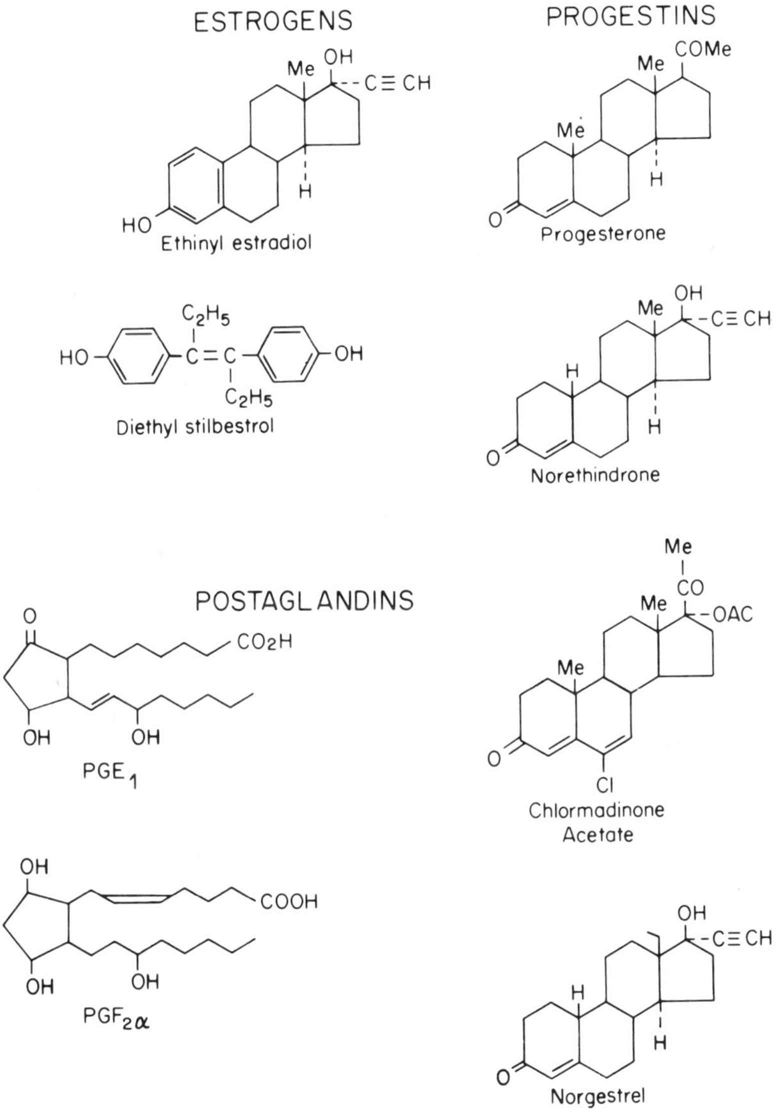

Figure 16. Chemical formulae of some estrogens, progestins and prostaglandins and common oral contraceptives, norethindrone, chlormadinone acetate and norgestrel. Note the similarities and differences between ethinyl estradiol, progesterone and the synthetic oral contraceptives.

implantation of the blastocyst, preparation of mammary development for milk production, and regulation of uterine contractions during parturition. Sex steroids also play a major role in the phenotypic expression of sex. The differentiation of male or female fetus depends on the ratios of testosterone to estrogen production during prenatal development.

The secretory activity of steroid hormones by the gonads is under endocrine control of the anterior pituitary. Hypophysectomy before or after puberty causes the gonads to atrophy. The secretory activity of the gonad is restored by injection of pituitary preparation or implantation of pituitary tissue.

Androgens

Testosterone, the principal circulating sex steroid in men, is secreted primarily by the Leydig cell of the testis. Circulating estradiol in men appears to be derived primarily from peripheral conversion of testosterone.

Estrogens

Estradiol is the biologically active estrogen produced by the ovary, with smaller quantities of estrone. Except for the possible secretion of small amounts of estriol in the luteal phase of the menstrual cycle, most estriol and related urinary estrogens are metabolic breakdown products of secreted estradiol/estrone. All estrogens secreted by the ovary are produced from androgenic precursors.

Progesterone

The corpus luteum produces progesterone as the dominant hormone, but also produces significant quantities of estradiol. Progesterone causes the typical secretory endometrium of the luteal phase of the menstrual cycle. Uterine motility under the influence of progesterone is characterized by spontaneous contractions of low frequency and high amplitude. Progesterone has little effect on the genital tract if administered alone, but has striking effects if given after estrogen.

FEEDBACK MECHANISMS

The synthesis, storage and release of hypothalamic hormones are regulated by both pituitary and steroid hormones through two feedback mechanisms: a long and a short loop. Long feedback involves interaction among the gonad, pituitary and hypothalamus. In the short feedback system, the levels of pituitary gonadotropins can influence the secretory activity of the releasing hormones without mediation of the gonads.

Depending on their concentration in the blood, steroid hormones may exert a stimulatory (positive) or an inhibitory (negative) feedback. Positive feedback results when an estrogen or a progestin stimulates the release of a gonadotropin, such as LH. Negative feedback results when large levels of progesterone prevent the release of a gonadotropin.

Stimulatory Feedback (a *Positive* Feedback)

In this system, an increasing level of hormones causes subsequent increase of another hormone. For example, increasing levels of estrogen during the preovulatory phase trigger an abrupt release of pituitary LH. These two events are precisely synchronized, since an LH surge is necessary for the rupture of the ovarian follicle.

Inhibitory Mechanism (a *Negative* Feedback)

This system involves reciprocal interrelationships with two or more glands and target organs. For example, as the ovary is stimulated, estrogen secretion increases, and FSH levels decline. Also, when pituitary hormones reach a certain level, some hypothalamic nuclei respond by decreasing the production of their particular releasing hormone. The decreased levels of releasing hormones causes a decline in secretion of pituitary tropic hormone and, subsequently, a lower level of target gland function.

PROSTAGLANDINS (PG)

Prostaglandins, derivatives of long-chain polyunsaturated fatty acids, are found in various tissues of many species ranging from corals to mammals. There are six primary, naturally occurring prostaglandins, but human reproduction is influenced by two main series–PGE and PGF.

Prostaglandins may be considered hormones, which regulate several physiological and pharmacological phenomena, *e.g.*, contraction of smooth muscles in the reproductive and gastrointestinal tracts, erection, ejaculation, sperm transport, ovulation, formation of the corpus luteum, parturition and milk ejection.

The human semen contains high concentrations of prostaglandins. Low levels of prostaglandins in semen may be related to male infertility since the concentration of prostaglandin E in the seminal plasma is about 18 μg of PGE/ml in infertile men and 55 μg/ml in fertile men. The intravenous infusion of PGE_2, and the intra-amniotic and extra-amniotic infustion of $PGF_{2\alpha}$ is an effective abortifacient. Several side effects may result from activation of smooth musculature of gastrointestinal tract, *e.g.*, nausea, vomiting, diarrhea and epigastric pain. Prostaglandins have been tested as a possible once-a-month contraceptive administered as a single vaginal suppository at the time of the expected menstrual period. Their contraceptive action may involve an effect on the corpus lutuem, causing it to regress; an effect on the myometrium, increasing uterine motility; or a direct effect on the endometrium, leading to its regression and induced menstruation.

HORMONE RECEPTORS

Each hormone has a selective effect on one or more target organs. This phenomenon is achieved through two mechanisms: (1) each target organ has a specific method of binding that hormone not found in other tissue; or (2) the target organs have certain metabolic pathways capable of responding to the hormone-metabolic pathways not shared by nontarget tissue. It seems that specific binding is the usual mechanism. For example, all target

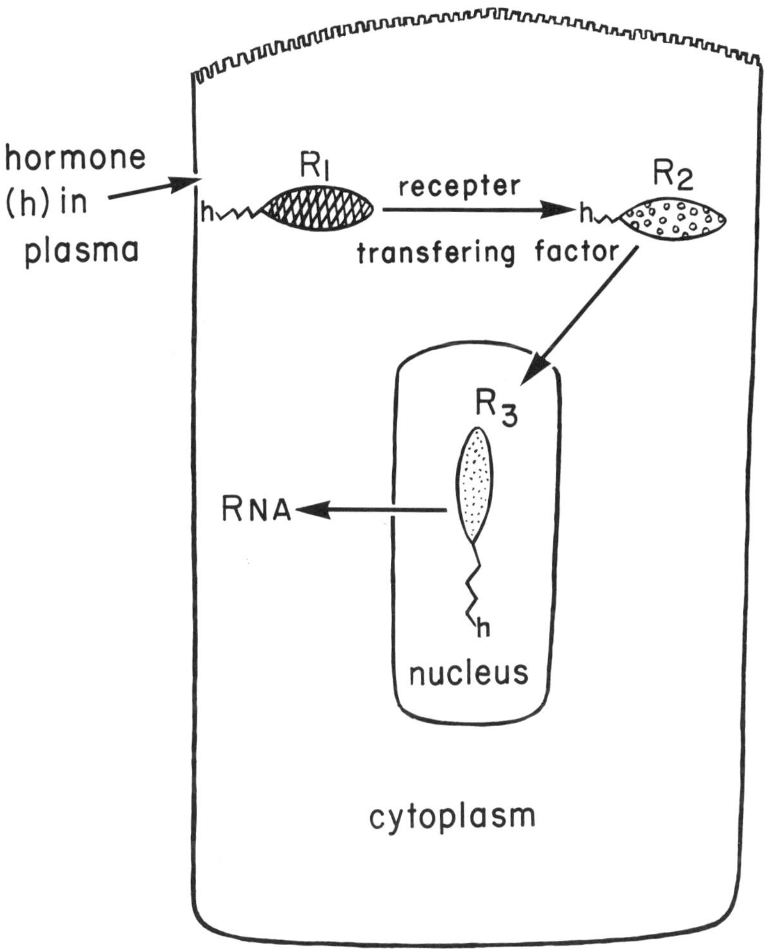

Figure 17. Simplified scheme to illustrate receptor theory and the mechanism of hormone activation to target tissue. The molecules of hormone in the plasma enter the cytoplasm of the cell of the target tissue and bind specific receptors (R_1) to that hormone. The receptors are then transformed to another type (R_2), which can enter the cell nucleus and be transferred to a third type (R_3), which exerts dramatic effects on RNA synthesis.

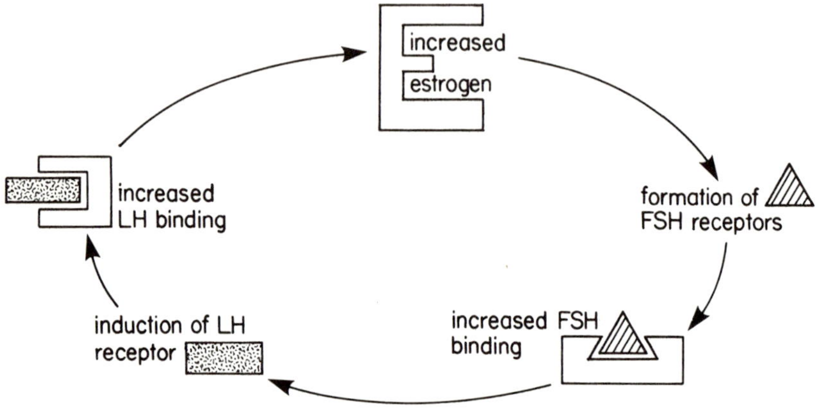

Figure 18. Diagram showing the formation of gonadotropin receptors as affected by estrogen level.

tissues that respond to steroid hormones contain a receptor protein within the cell, which specifically binds the activating hormone.

Within the target cell, the steroid hormone is found in the cytoplasm, bound to a relatively large protein (molecular weight = 200,000). This steroid protein complex migrates across the cytoplasm close to the nucleus where, through an enzymatic process, the protein changes in shape and size (Figure 17). The newly transformed complex enters the nucleus and causes a sequence of physiological responses specific for that cell.

The target cells of the anterior pituitary possess cell membrane receptors which recognize and selectively bind the protein hormones, including gonadotropins. The binding phenomenon triggers the synthesis and secretion of the pituitary hormone via the cyclic AMP-protein kinase system of the cell. Gonadotropin receptors, in turn, are influenced by estrogen levels (Figure 18).

SECTION II

FUNDAMENTALS OF HUMAN REPRODUCTION

CHAPTER 4

SEXUAL MATURITY

Sexual maturity is the transitional period between the prepubertal state of sexual development and adult sexual maturation with attainment of reproductive capability. Sexual maturity is associated with several physiological responses: morphological and functional changes in the gonads; functional interactions among the hypothalamus, pituitary and gonads; growth spurt; developmental changes in the target organs and secondary sexual characteristics; emotional, behavioral and psychosexual changes. The sequence of events (Table VII) is of considerable clinical significance in the evaluation of the rate and stage of sexual maturity.

Age and body weight at sexual maturity are influenced by several factors, such as heredity, nutrition, geographical location, climate, socioeconomic factors and the incidence of systemic diseases of childhood. Sex-determined morphological characteristics (sexual dimorphism) associated with puberty have physiological and behavioral significance in reproduction. In the female, a wide pelvis is required for parturition and well-developed mammary glands for lactation.

SEXUAL MATURITY IN BOYS

The physiological events associated with sexual maturity occur in the following sequence: increased testicular growth in response to increased pituitary activity; appearance of mitotic figures in the seminiferous tubules; onset of spermatogenesis; testosterone excretion in the urine; and changes in the larynx,

Table VII

Sequence of Physiological Events Associated with Sexual Maturity in American Girls and Boys

Sex	*Major Physiological Events*	*Age (yr)*
Girls	Bony pelvis grows and widens	9-10
	Female contour assumed and fat deposition	
	Budding of nipples	10-11
	Budding of breasts	
	Appearance of pubic hair; in 10% of girls, pubic hair appears before hypertrophy of breasts	
	Vaginal secretion	
	Glycogen content increases in vaginal epithelium, pH of vaginal secretions becomes acidic	11-12
	Pigmentation of nipples	
	Growth of breasts	12-13
	Axillary hair appears	
	Areola appears above plane of breast	
	Menarche occurs	13-14
	Pubic hair increases in amount	
Boys	Leydig cells multiply and androgen secretion increases	9-10
	Mitotic activity appears in seminiferous tubules	10
	Active spermatogenesis begins	13
	Testicular length 4-5 cm; darkening and enlargement of scrotum; penis increases in length and diameter	13.8
	First ejaculation	11-15
	Testicular length 5 cm; pigmentation and thickening of scrotum; spermatogenesis and semen characteristics reach adult levels	

skin and body hair distribution. Although there is a gradual increase in testicular size with age after birth, this is only in proportion to the increase in the size of other organs with general body growth. However, the onset of puberty is associated with dramatic changes in the size and histology of the testes (Table VIII) and in the levels of serum testosterone. The seminiferous tubules become more tortuous and acquire large lumina. Along

Table VIII

Changes in Structure of the Testis from Birth to Senescence

Age (yr)	*Structural and Functional Changes in Testes*
At Birth	Seminiferous tubules 70 μ in diameter, filled with undifferentiated cells with no lumen
4	Lumen develops in seminiferous tubules without differentiation in germinal epithelium
10	Pituitary gonadotropins cause rapid enlargement of seminiferous tubules; Spermatocytes differentiate; Leydig cells reappear
13-16	Spermatogenic development reaches adult stage; some 500 seminiferous tubules are present
Adult	Seminiferous tubules, 150-250 μ in diameter
Aged Men	Spermatogenic activity declines in different men at variable ages (50-70 years) Actively functioning tubules decrease in number Sclerosis, peritubular fibrosis and premature sloughing of spermatogenic epithelium

with the changes in the tubules, mitotic activity in the germinal epithelium begins. The age at which spermatogenesis is initiated is not easy to estimate since it may precede capacity to ejaculate. However, boys usually produce fertile sperm at 14-15 years of age.

In response to increased levels of testosterone at adolescence, the penis enlarges and the scrotum develops rugae. The wide variability in penile size raises difficult social and educational problems and may cause psychological maladjustments. The acceleration of the growth of the larynx does not occur until the termination of the penile growth spurt, and is presumably due to direct stimulation of testosterone by the cells of laryngeal cartilages and the associated soft parts.

Puberty is associated with morphological and functional changes in the skin and hair. Young boys have darker skin on the chest and forehead and, to a greater extent, a darker mammary areola than young, nulligravida girls. Increased secretory activity in sebaceous glands of the scrotum, face, back and chest may cause perspiration of a characteristic odor. Pubic hair appears before or after onset of body height spurt, whereas facial and axillary hair appear two years after the first growth of pubic hair. Increased oiliness of the skin and penile erection are more sensitive criteria of androgen than the growth of pubic hair.

SEXUAL MATURITY IN GIRLS (MENARCHE)

Menarche, the onset of menstrual bleeding, is a dramatic event which is more easily remembered than the onset of puberty in boys. The first several menstrual cycles are usually irregular and anovulatory. Regular ovulatory cycles are usually established within two years after menarche, and full fertility is attained a few years after the first menstruation (Figure 19).

Secondary sexual characteristics develop in the following sequence: breast growth, pubic hair, axillary hair and menstruation (Figure 20). Tanner classified breast development into five stages:

1. elevation of papilla only during preadolescence;
2. elevation of breast and papilla as a small mound, with increased areolar diameter;
3. further enlargement and elevation of breast and areola, with no separation of their contours;
4. projection of areola and papilla to form a secondary mound above the level of the breast; and
5. projection of papilla only, owing to recession of the areola to general breast contour during the mature stage.

Increased levels of circulating ovarian estrogens cause enlargement of the uterus and increased response of the endometrium. Estrogens also cause pigmentation of the areola, growth and mobility of the nipple, and growth and possible pigmentation of the vulva. Axillary and pubic hair appear in response to alteration of adrenal function (adrenarche) and are lost in cases of severe adrenal insufficiency.

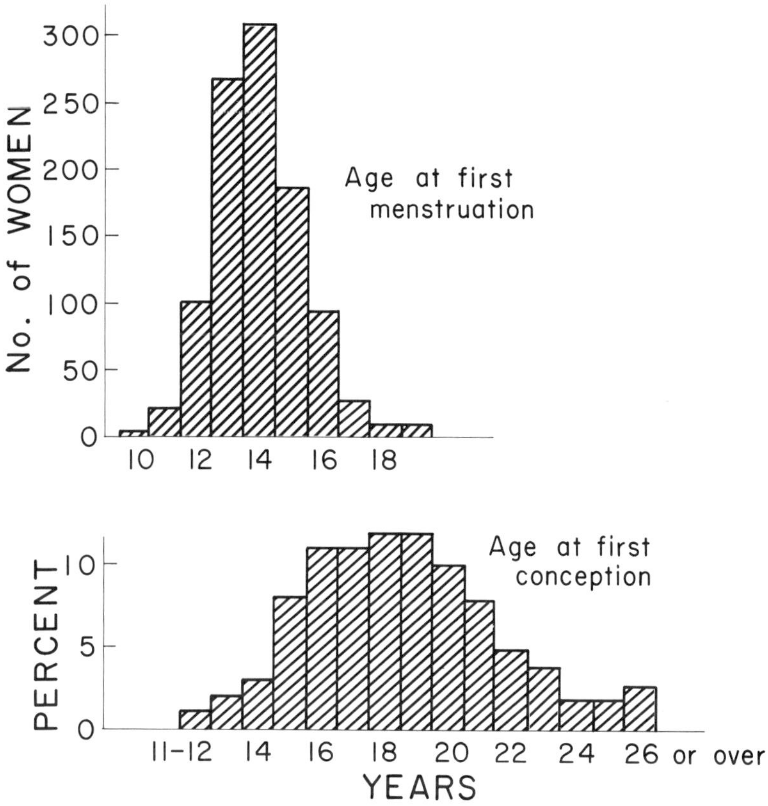

Figure 19. Distribution frequency of age at first menstruation and age at first conception, in Indian women (From Peters, H. and Shrikande S. M., *Fertil Steril,* 8:355 1957.)

BODY GROWTH SPURT AND BODY COMPOSITION

The adolescent spurt of body growth occurs two years earlier in girls than in boys, but is somewhat less marked (Figure 21). Regardless of age of menarche, American girls grow 4 to 10 cm (average 6 cm) after menarche. The age at which growth ceases is more closely related to the age of adolescence than to chronological age. Thus, if the height spurt occurs early in a boy, growth of the genitalia, increase in muscular strength and

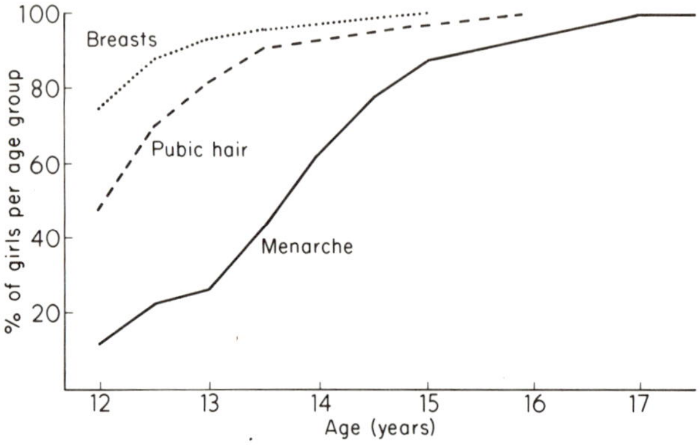

Figure 20. Sequence of physiological events associated with puberty in girls (From Land, G. M. V'ant, and J. H. deHaas,: *Ned Tijdschr Geneesk* 101:1425, 1957; Donovan, B. T., and J. J. van der Werff ten Bosch,: *Physiology of Puberty,* Arnold, London, 1965.)

attainment of adult stature also occur early. Dietary and socio-economic factors in the childhood environment play a major role in somatic development as measured by body height and weight. For example, chronic undernutrition causes a decline in the rate of skeletal growth and sexual maturation.

The total muscle mass in the body can be estimated by several methods such as dynamometry, roentgenometry, creatinine determination and body density. During adolescence, different growth patterns of muscle, fat and bone in boys and girls, cause remarkable variations of their body composition. By age 16, the volume and strength of muscle are greater in boys than in girls, and by their late teens, the difference is twofold. A steady increase in total body fat occurs as adolescence progresses. However, as age increases during adolescence, subcutaneous fat increases in girls and decreases in boys. These changes in body composition during adolescence are more related to endocrine factors than to food intake. The smaller size of the female and the characteristic distribution of subcutaneous fat which causes her rounded contours are partly caused by the bone-ossifying and fat-mobilizing effects of estrogen, whereas greater muscular mass in boys is due to the anabolic action of androgens.

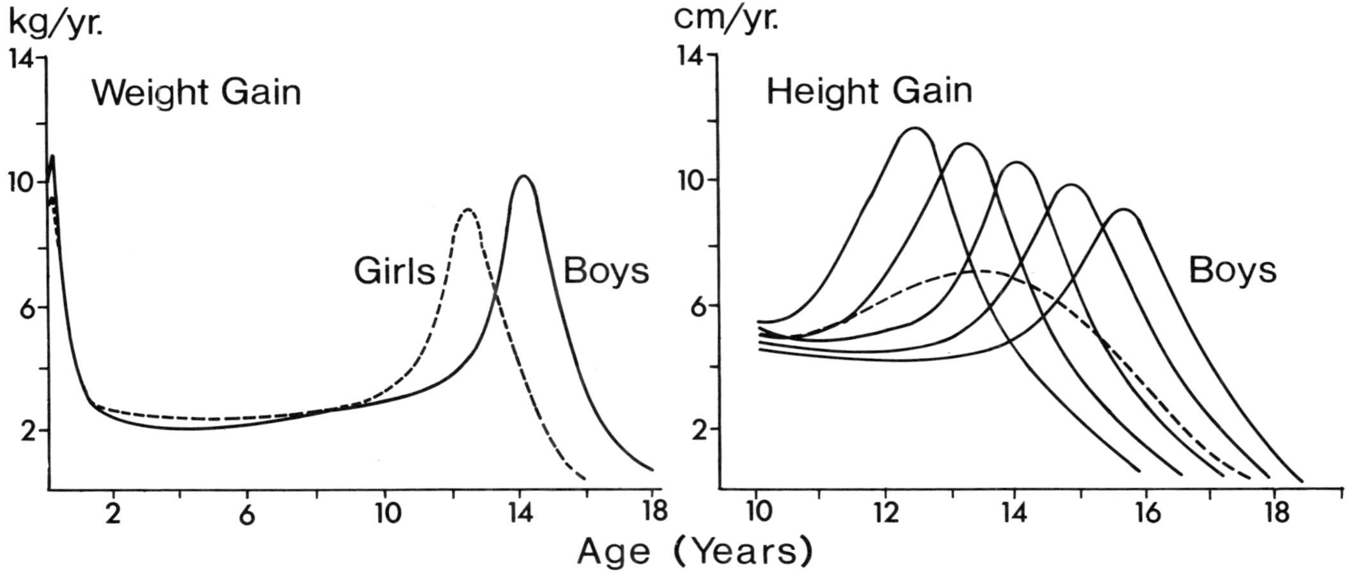

Figure 21. *Left.* Typical individual velocity curves for weight: English boys and girls. *Right.* Adolescent spurts in height. Individual velocity curves of five boys. The dashed curve is obtained by averaging individual values at each age. Note that average curve does not show the dramatic nature of the spurt in the individual case (From Tanner, J. M., *et al. Arch Dis Child* 41:454, 1966; Sinclair, D. C. *Human Growth After Birth*, Oxford, 1969.)

ENDOCRINE CONTROL

Sexual maturity is regulated by complex physiological interaction of the hypothalamus, pituitary, gonads and sexual accessory organs:

1. Maturational changes in hypothalamic stimulation are associated with decreased feedback sensitivity to steroids.
2. There is increased sensitivity of the anterior pituitary to stimulation by hypothalamic gonadotropin-releasing hormone (GnRH).
3. There is increased response of the testis and ovaries to FSH/LH stimulation.
4. There is increased sensitivity of sexual accessory organs to stimulation by gonadal steroids.

At birth, the infantile gonads may secrete small amounts of gonadal hormones which inhibit the hypothalamus from producing GnRH. Gradually, gonadotropins are secreted at a very low level, stimulating gonadal hormone secretion at a low level. A sudden rise of gonadotropins at puberty stimulates the gonads. Meanwhile, the gonadal hormones exert inhibitory effects on the hypothalamus which secondarily modifies pituitary function. Rising levels of FSH cause the differentiation of seminiferous tubules of the testis, and the follicles of the ovary. Increased levels of LH cause the Leydig cells to secrete testosterone, and cause rupture of the ovarian follicle.

In adolescent girls, cyclic excretion of estrogens precedes menarche, breast development, and the growth of pubic hair. Increased estrogen levels, during the menarcheal stage, cause the maturation of anovulatory follicles. Before FSH and LH can trigger ovulation, the ovarian follicles must develop a "competence" to respond to stimulation by gonadotropins. This "competence" is generally characteristic of large antral follicles.

In adolescent boys, androgens perform various functions:

1. stimulate development of the penis, scrotum, seminal vesicle and prostate;
2. promote growth of the beard and development of the larynx;
3. exert effects on the distribution of body hair, body fat,

muscle mass, bone growth and epiphyseal closure; and
4. stimulate development of sexual desire.

Androgen secretion by the testis is directly regulated by circulating levels of pituitary LH which, in turn, is controlled by gonadotropin-releasing hormone (GnRH) from the hypothalamus.

The development of secondary sexual characteristics results from increased testicular secretion of testosterone in boys or increased ovarian secretion of estrogens in girls.

ADOLESCENT FERTILITY

Adolescent menstrual cycles are usually anovulatory, lacking a functional corpus luteum. Because of the absence of progesterone dominance, menstruation is not associated with high levels of myometrial activity that are found when the progesterone is withdrawn and, therefore, dysmenorrhea is infrequent. Seasonal oligoamenorrhea, especially during the summer months, may persist in certain adolescent girls.

The period of adolescent infertility may last 1 to 2 years after menarche, and regular ovulation is not established until several years after menstruation has begun. The period of adolescent infertility enables the girl to attain greater maturity before reproduction. This is particularly advantageous since behavioral maturity may lag behind physical development. Little is known about comparable infertility in adolescent boys. Adolescent marriages result in few pregnancies during their initial years, despite the high frequency of intercourse. Despite this relative infertility, pregnancy can occur at an early age. Since the age of menarche is declining in some populations, a large proportion of teenage girls can now conceive at an earlier age. Adolescent pregnancies are associated with increased incidence of toxemia, uterine inertia, contracted pelvis, prolonged labor, prenatal mortality, fetal anomalies or premature labor.

Table IX

Some Causes of Delayed Puberty in Boys and Girls

Organ and Parameters	*Causes and Examples*
Hypothalamus	Congenital absence or deficiency in hypothalamic gonadotropin-releasing hormones (GnRH) Hypothalamic cyst or tumor Suppression by exogenous gonadotropins or steroids Psychological trauma
Pituitary	Congenital aplasia or hypoplasia Vascular deficiency Carcinoma, granuloma, trauma
Ovary-Testis	Ovarian or testicular dysgenesis Trauma, torsion, tumor Enzyme deficiency, neoplasm Orchitis-mumps, orchietomy
Endocrine Glands	Hyperthyroidism, hypothyroidism Thyroiditis, hypoadrenalism
Reproductive Tract	Imperforate hymen Congenital absence of uterus and/or vagina Congenital absence of epididymis
Chromosomal and Congenital	True hermaphroditism, pseudohermaphroditism Turner's syndrome Testicular feminization
Nutritional	Malnutrition, zinc deficiency Anorexia nervosa
Systemic Diseases	Rheumatoid arthritis, asthma, ulcerative colitis, enteritis, renal tubal acidosis, severe diabetes mellitus, gonorrhea, tuberculosis, sickle cell anemia
Other	Radiation, toxins

DELAYED AND PRECOCIOUS PUBERTY

Normal pubertal development lasts for 4 to 5 years. Puberty is delayed if physical changes are delayed beyond age 13 in girls and age 14 in boys, or if genital growth is not completed within 5 years from the time of the first physical signs of puberty or menarche. Delayed puberty may be due to anomalies in the hypothalamic-pituitary axis, decreased levels of gonadotropins, insensitivity of target organs to gonadotropins, congenital malformation, or absence of a reproductive organ and gonadal dysgenesis (Table IX). Primary hypogonadism results in a thin, flat-chested, rapidly growing adolescent girl with delayed or absent menarche and hypoplastic genitalia.

The ovaries of prepubertal girls are fully prepared to function if they receive the appropriate gonadotropic stimuli. Precocious puberty—the onsent of menstrual cycles considerably before the age of 10—may be idiopathic or may be secondary to several lesions in the central nervous system. A well-known case of precocious puberty is that of Lina Medina, who began menstruation at eight months of age and gave birth to a child by Cesarean section at the age of five and one-half years. This was due to Albright's syndrome, characterized by premature development of the body and sex characteristics, with or without precocious menstruation.

Anomalies of sexual maturity in boys include delayed puberty, adolescent hypogonadism, isosexual precocious puberty and pseudoprecocious puberty. Adolescent hypogonadism in boys is caused by primary or secondary testicular failure or androgen insufficiency. Isosexual precocious puberty due to hypothalamic malfunction with premature release of GnRH and pituitary gonadotropins leads to testicular androgen production and spermatogenesis. Brain lesions may be structurally demonstrable or idiopathic. Precocious puberty is also caused by androgens derived from testicular or adrenal tissues as a result of adenoma or carcinoma or to loss of endocrine regulatory mechanisms.

CHAPTER 5

FUNDAMENTALS OF ANDROLOGY

Reproductive ability of the male is regulated by hereditary factors, prenatal differentiation, endocrine milieu and pubertal development. The ability to produce viable spermatozoa capable of fertilization depends on various physiological parameters:

1. the presence of a normal responsive hypothalamus, pituitary, testis and accessory glands;
2. a balanced and coordinated function of hypothalamic-releasing hormones, gonadotropins and testicular hormones;
3. active spermatogenesis (qualitative and quantitative);
4. patency of excurrent ducts permitting transport of sperm through epididymis, vas deferens and ejaculatory duct;
5. secretory function of accessory sex glands; and
6. normal vascular, neurological and psychological components of erection and ejaculation.

ERECTION AND EJACULATION

Erection

Erection of the penis, primarily a vascular phenomenon, is initiated by a complex nerve pathway involving the cerebral cortex. Erection may also be initiated by purely psychic stimuli, but its integration takes place in the spinal cord sacral segments. Precoital stimulation causes remarkable dilation of the arteries supplying the cavernous bodies of the penis. The increased pressure in the cavernous spaces of the corpus cavernosum causes the protrusion of the penis. The blood is trapped in the erectile tissue during sexual excitement.

Ejaculation

Ejaculation involves complex motor activity which is completed in two phases: emission of sperm and ejaculation of semen.

Emission of Semen

Sperm from the epididymis and vas deferens as well as secretions from the seminal vesicles and prostate are transported into the posterior portion of the urethra by active contraction of the musculature around the epididymis, vas deferens and seminal vesicle. Emission is associated with contraction of the internal urethral sphincter and closure of the bladder neck, thus preventing the ejection of the sperm backward into the bladder.

Ejaculation of Semen

This phase seems to be triggered by the arrival of semen in the prostatic part of the urethra. Afferent impulses stimulate the sacral and lumbar nerves which, in turn, stimulate spasmodic contraction of bulbocavernosus and perineal muscles. The external urethral sphincter relaxes and semen is propelled via the urethra through the external meatus. Semen is ejaculated in several forceful thrusts. Disorders in the ejaculation phase may lead to "retrograde ejaculation" of semen into the urenary bladder.

Ejaculation involves specific synchronizing mechanisms. Thus, the accessory sexual glands do not expel their respective secretions simultaneously during the emission phase. The secretions of the prostate and bulbourethral glands appear in the first portion of the ejaculate, whereas the final part consists almost entirely of secretions of the seminal vesicle.

The emission of sperm is controlled by the hypogastric nerve, whereas ejaculation of semen is regulated by the parasympathetic internal pudendal nerve. The importance of an intact adrenergic transmission in the vas deferens for ejaculation is substantiated by the failure of emission noted in patients receiving adrenergic blocking substances, *e.g.*, guanethidine.

PHYSIOLOGY OF SEMEN

The ejaculate consists of a mixture of sperm and seminal plasma made up of various secretions from the male accessory glands. Semen is ejaculated on the external opening of the cervix and posterior vaginal fornix. The ejaculate coagulates within one minute, trapping most sperm. Several enzymes of the seminal plasma cause complete or partial liquefaction of the coagulum within 5 to 20 minutes. Sperm motility is not fully attained until liquefaction of the coagulum. The physiological significance of semen coagulation for sperm transport is not known. Men whose semen liquifies slowly, *i.e.*, two to three hours or longer, are therefore frequently subfertile. The pH of the vaginal fluid is usually acidic. However, the cervical secretions which coat the upper part of the vaginal walls considerably increase the alkalinity of the vaginal milieu and provide a favorable medium for sperm activity.

During intercourse, some 200 to 400 million sperm are ejaculated. However, the first fraction of the ejaculate contains the highest concentration of sperm. The number of sperm in the ejaculate is an important factor for fertility, but the minimal number of sperm necessary to insure fertilization is not known. Unfortunately, there is no precise information on the normal range of semen characteristics with respect to fertility based on sperm concentration, total sperm count, motile activity, and the proportion of motile sperm in the ejaculate. The normal, healthy fertile semen contains between 50 and 150 million sperm/ml; in excess of 60 or 70% of these are actively motile (Table X). Some examples of semen analysis in relation to male infertility include:

1. semen with average volume, concentration 100 million/ml; 80% forward motility (in one direction); definitely fertile;
2. semen with 10 million/ml, less than 10% motility; definitely infertile; and
3. semen with sperm concentration between 20 and 40 million/ml; 40-50% of average or indifferent motility; fertility in dispute.

Table X

Characteristics of Human Semen of Fertile Men

Characteristics	*Values*	
	Range	*Mean*
Volume of Ejaculate, ml	1-6	3.5
Sperm Concentration, million/ml	50-150	90
Sperm Count, million/ejaculate	100-500	300
Coagulation-Liquefaction	Coagulates upon ejaculation then liquefies in 5-20 minutes	
Color and Odor	Translucent to whitish-grey, smells like chestnut flowers	

Seminal Plasma

The seminal plasma comes from secretions of the epididymis, ampulla, seminal vesicles, prostate, bulbourethral and urethral glands. The seminal plasma contains various ions, amino acids, lipids, carbohydrates, proteins, hormones and enzymes. Specific components of the secretions of the prostate (*e.g.*, acid phosphatase, zinc, manganese) and the seminal vesicles (*e.g.*, fructose, prostaglandins) are utilized as biochemical markers for these glands. Biochemical analyses of seminal plasma are frequently undertaken routinely in the evaluation of male infertility. These analyses may serve as a useful guide to regions in the reproductive tract that are sites of dysfunction. Thus, measurement of the citric acid or acid phosphatase concentrations reflect the adequacy of prostatic secretions while the measurement of fructose concentration is an index of the secretory activity of the seminal vesicles. It is probable that the role of fructose is to provide energy for spermatozoa. At the time of ejaculation, when sperm come into contact with secretions that are rich in fructose, they become highly motile, an activity for which energy is required.

Seminal plasma plays a major role in the transport and the physiology of sperm. However, sperm from the vas deferens and epididymis are successfully used in artificial insemination, and removal of various accessory organs of the male tract rarely decreases fertility as long as ejaculation still results in the release of a few million sperm. Different secretions of the accessory glands have different effects on motility, survival and metabolism of spermatozoa. The prostatic secretions have a promoting and protecting effect, whereas the vesicular secretions exert a harmful effect.

The seminal plasma contains steroids and gonadotropins. Steroids seem to originate either from Leydig cells' secretion followed by diffusion to the tubules, or from tubular biosynthesis. It is also possible that the steroids of the seminal plasma may be secreted from the seminal vesicle or prostate. In the seminal plasma, LH concentration is higher than in blood plasma, whereas FSH in the seminal plasma is lower than in blood plasma.

Sperm

Sperm normally constitute no more than 5% of the volume of the human ejaculate.

Morphological Characteristics

The sperm is made of a head, middle piece, and tail (Figure 22). The head is flattened and contains condensed genetic material. The acrosome covers the anterior one-half to two-thirds of the sperm nucleus. Unlike other mammals, human sperm show a greater variability in head size and shape.

Several types of immature germ cells appear in the semen, *e.g.*, spermatids, spermatocytes, and spermatogonia. A count of more than one million spermatids/ml is indicative of a serious spermatogenic derangement and may result in male infertility. The presence of large numbers of sperm with cytoplasmic droplets indicates epididymal dysfunction or that the patient was abstinent for a very short period of time. The latter may be determined by asking the patient to be abstinent for a longer interval and repeating his semen analysis.

Sperm may exhibit different types of morphological abnormalities, *e.g.*, tapered, giant, pin, double, quadruple or frazzle-headed with coiled, bent or two tails. Certain infertile men

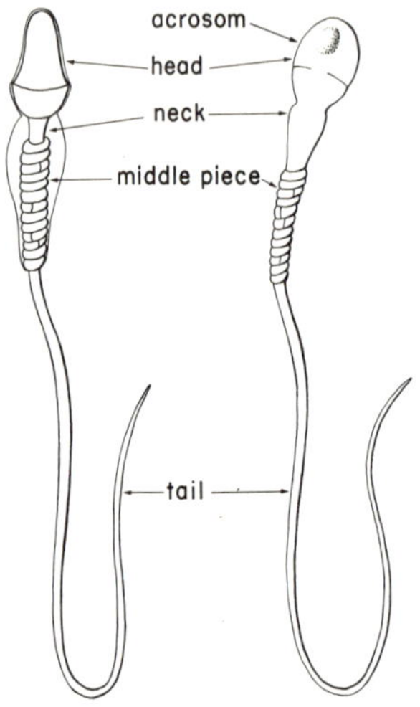

Figure 22. Schematic drawings of human sperm from front (right) and lateral (left) views. The nucleus is located in the head, whereas the spiral mitochondria are located in the middle piece (From Pedersen, H. and D. W. Fawcett, *Human Semen and Fertility Regulation in Men.* E. S. E. Hafez, Ed., Mosley, St. Louis, 1976.)

produce sperm with rounded, globular heads, whereas the sperm count and motility are within the normal range.

X- and Y-containing sperm do not seem to vary in shape and dimension. Several unsuccessful attempts have been made to control sex ratio by separating X and Y sperm using various experimental techniques, but no conclusive results have been obtained. The long arm of the Y chromosome in somatic cells exhibits an intense fluorescence when stained with certain fluorescent compounds. The same technique is used to identify Y and X human sperm and can be readily distinguished by the presence in the former of a discrete fluorescent spot in the Y sperm.

Physiological Aspects

There are several types of sperm motility, but two types are common: (1) directional progressive motility, in which the sperm actually moves forward in one direction; and (2) vibratory motility, in which the sperm moves its tail, but does not progress in a forward direction. Since motility rating is higher in the first half of the ejaculate, it is used for artificial insemination.

The patterns of sperm motility change as sperm encounter the changing ionic milieu and biophysical states of the epididymal fluid, seminal plasma, cervical mucus, endometrial fluid, oviductal fluid and periotoneal fluid. The acidic vaginal environment is hostile to sperm survival. Sperm transport through the cervix is affected by contractile activity of vagina and cervix, properties of cervical mucus, rate of liquefaction of coagulum, directional motility of sperm and possibly orgasm. Sperm penetration in cervical mucus depends on its biophysical and biochemical characteristics, pH, degree of cellularity, and presence of immunoglobulins in the mucus, proteolytic enzymes in sperm, and concentration of prostaglandin and related substances in seminal plasma. During their transport in the female tract, sperm are separated from seminal plasma and resuspended in endometrial and oviductal fluid. Sperm velocity increases in the presence of oviductal or follicular fluid. This increased motility may be of physiological significance during fertilization of the egg.

HORMONAL CONTROL OF MALE FERTILITY

Sexual activity and spermatogenesis are regulated by interactions between the central nervous system, hypothalamus, pituitary and testes (Figure 23). Androgens are also involved in regulating sexual behavior, maintenance of the secondary sexual characteristics and feedback on the hypothalamo-pituitary axis to control secretion of gonadotropins.

Androgens

Androgens are released in high concentration from Leydig cells, directly into the seminiferous epithelium. This pattern has an important direct role, not only on the maintenance of spermatogenesis, but also on the physiological integrity of the epididymis and the survival of epididymal sperm. The secretion of androgen from Leydig cells is regulated by the output of LH

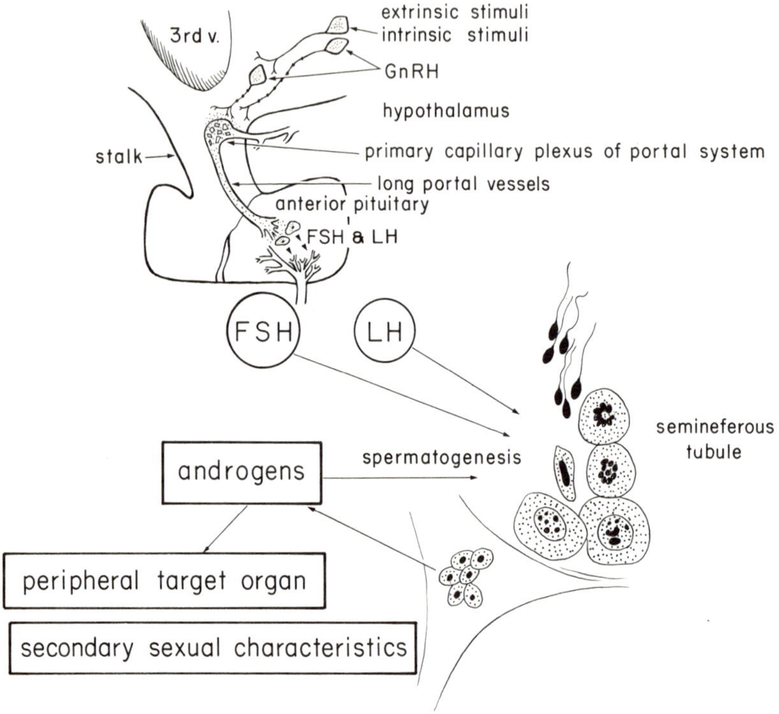

Figure 23. Endocrine control of male reproduction. Note the complex interactions in the hypothalamus-pituitary-testes axis. Androgens produced by Leydig cells (in interstitial tissues) are necessary for spermatogenesis and expression of secondary male sexual characteristics.

from the anterior pituitary which, in turn, is controlled by hypothalamic gonadotropin-releasing hormone (GnRH). Testosterone in blood plasma is mainly bound to plasma proteins, only a small fraction being free.

Gonadotropins

Pituitary gonadotropins are essential for normal testicular function. Removal of the pituitary (hypophysectomy) causes cessation of spermatogenesis and atrophy of the testes and accessory sex organs. Treatment of hypophysectomized males with pituitary extracts restores testicular function.

LH regulates the secretion of testosterone by the Leydig cells, whereas FSH acts on Sertoli cells and is necessary for

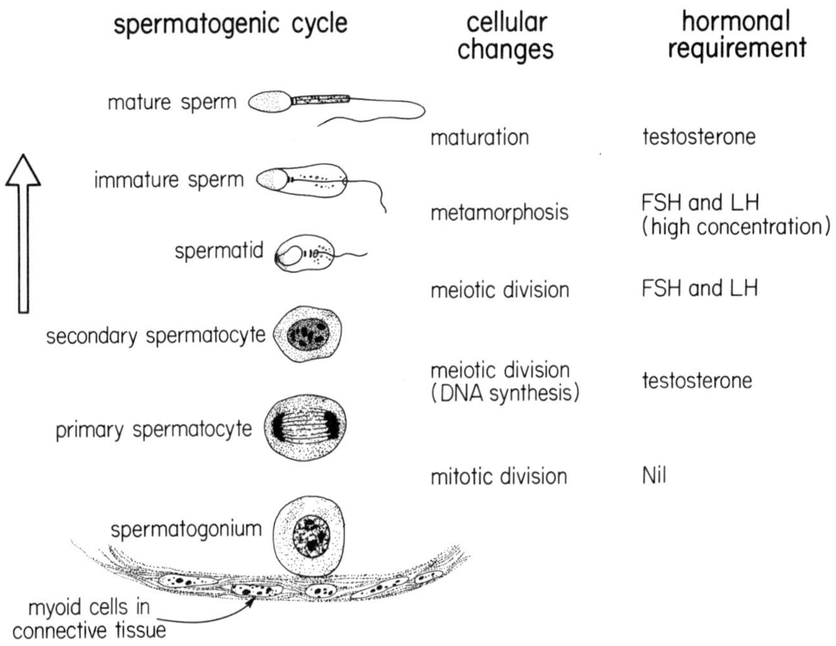

Figure 24. Spermatogenic cycle showing cellular changes within the seminiferous tubules of the testes. Note cytological changes in the nucleus from spermatogonium to primary spermatocyte, secondary spermatocyte spermatid, immature and mature sperm. Various maturational phases in the spermatogenic cycle have different hormonal requirements.

initiation of spermatogenesis (Figure 24). Once spermatogenesis is initiated, its maintenance probably depends on the combined effects of a high local concentration of androgens from the Leydig cells and possibly the Sertoli cells. Androgens seem to stimulate the proliferation of spermatogonia which, in turn, mature to secondary spermatids. However, the final stages of spermatogenesis require RSH.

Hypothalamic Gonadotropin-Releasing Hormone (GnRH)

Intrinsic and extrinsic stimuli exert specific effects on centers in the hypophysiotrophic area of the hypothalamus. These contain the GnRH-producing neurons, whose axons end at the capillary loops of the hypophysical portal circulation. Hypothalamic gonadotropin-releasing hormones (GnRH) produced by these neurons are discharged into the protal system of the pituitary gland where secretion of gonadotropins is regulated. Testicular steroids inhibit hypothalamic GnRH secretion, whereas androgens diminish pituitary responsiveness to GnRH with respect to LH release but exert no effect on FSH.

Inhibin

It has been suggested that a seminiferous tubule factor, referred to as "inhibin," is secreted by the Sertoli cells and excreted in the fluid of the rete testis and seminal plasma. It is believed that this substance is a polypeptide or protein in nature. Inhibin inhibits FSH secretion by modulating pituitary responsiveness to GnRH.

Feedback Mechanisms

The effects of LH are probably mediated by the stimulation of increased concentrations of intratesticular testosterone. Serum gonadotropin concentrations are, in turn, controlled by androgens of the testis. Testosterone exerts a negative feedback effect on LH and FSH secretion. Removal of these feedback effects, such as occurs after castration, results in increased LH and FSH levels.

A negative feedback may result from decreased output of GnRH or a reduction in the ability of the pituitary gland to respond to GnRH. Appropriate concentrations of androgens may result in a positive feedback, which causes a remarkable increase in gonadotropin levels. this may be the result of increased output of GnRH or increased ability of the anterior pituitary to respond to GnRH.

CHAPTER 6

MENSTRUAL CYCLE

Unlike man, rodents and farm animals experience estrous cycles characterized by short periods of "estrus" during which the female permits mating; ovulàtion takes place either shortly before or shortly after estrus. Human and nonhuman primates have defined peaks of sexual desire and, as a rule, permit intercourse throughout the entire cycle.

Reproductive life in women lasts some 30% of their life span. With the establishment of the menstrual cycles and the correct hormonal balance at menarche, follicular growth culminates in ovulation.

MENSTRUAL BLEEDING AND CYCLE LENGTH

Menstruation is the periodic physiological bleeding from the endometrium. Menstrual bleeding is comprised of red blood cells and leukocytes, desquamated endometrium, cervical mucus and vaginal cells (Table XI). Menstrual blood clots in the endometrial cavity, but the clots are lysed by fibrinolytic enzymes. The blood is discharged in a fluid state, and normally does not clot, but may clot with heavy flow.

Menstrual flow is preceded by a thinning of the endometrium, vasoconstriction, and increased coiling of the spiral uterine arteries. These phenomena lead to local necrosis and to desquamation of the superficial endometrial layers. With the withdrawal of progesterone from the plasma, the endometrium becomes thinner, the spiral arteries become exposed and bleeding

Table XI

Characteristics of the Human Menstrual Cycle

Parameter	*Values*	
	Range	*Mean*
Interval between menstrual cycles, days	24-32	28-30
Volume of blood loss per menstrual cycle, ml	3-7 50-75	5 60
Incidence of ovulation, days	14 ± 2 preceding onset of menstruation 8-32 hr after LH surge	
Constituents of menstrual blood	Blood = 50-75% Cervical mucus Desquamated endometrial tissue Vaginal cells	

begins. It is also possible that the absence of progesterone causes vasoconstriction, which leads to local ischemia, subsequent breakdown of the endometrium and menstruation.

Almost all women at some time experience dysmenorrhea (premenstrual tension and uterine cramps). These symptoms seem to be associated with ovulation and development of the corpus luteum, since anovulatory cycles, whether spontaneous or induced by hormone administration, are usually asymptomatic. The premenstrual stage may be associated with psychological irritability, depression and emotional instability.

Each woman's menstrual cycle has an individual pattern of length, duration, intensity and variability. The length of the menstrual cycle is traditionally a lunar month, but only 10-15% are exactly 28 days in length; the range is 24-35 days. Irregularities in the length of the menstrual cycles are caused by age, abrupt changes in climate and serious emotional disturbances. Maximal variation in the length of the menstrual cycle occurs during the first five years after menarche and the last six years before menopause. These two transitional periods, characterized by very short and very long menstrual cycles, vary widely in length and are very much like mirror images of one another. Unfortunately, there is no definite criterion for ending of the menarcheal or beginning of the menopausal transition. Minimal variation in the length of the menstrual cycle occurs around 35 years of age.

Cyclic menstrual bleeding, which continues more or less from 13-50 years of age, is interrupted only by pregnancy and subsequent lactation. Ovulatory menstrual cycles are resumed after various intervals, following parturition or abortion. Menstrual disorders following insertion of intrauterine devices may be related to the initial trauma and to biochemical and physical reactions of the uterus to a foreign body.

CYCLICAL CHANGES IN REPRODUCTIVE TRACT

Fluctuating levels of estrogen and progesterone throughout the menstrual cycle cause remarkable changes in the structure and secretions of the uterus, cervix and vagina (Figure 25).

Endometrium

The endometrium undergoes remarkable histological and cytological changes, which culminate in menstrual bleeding when the corpus luteum ceases to secrete its hormones. Much of the endometrial epithelium is shed at each menstrual bleeding. Following bleeding (postmenstrual phase), the endometrium enters a resting phase: the endometrium is thin, the endometrial glands are narrow and straight with a low columnar epithelium, and the endometrial stroma is rather dense (Figure 26). With rising levels of estrogen, the endometrium undergoes progressive proliferation and extensive mitoses (Table XII). Following restoration of the integrity of the endometrium, the cells appear poorly differentiated except for those lining the deepest endometrial glands.

The endometrial secretion, which is under endocrine control, provides an optimal environment for the transport and capacitation of sperm and the nutrition of the preimplantation blastocyst. If the ovum is fertilized, the endometrium continues to grow and becomes the decidua of pregnancy. The stromal cells swell, and

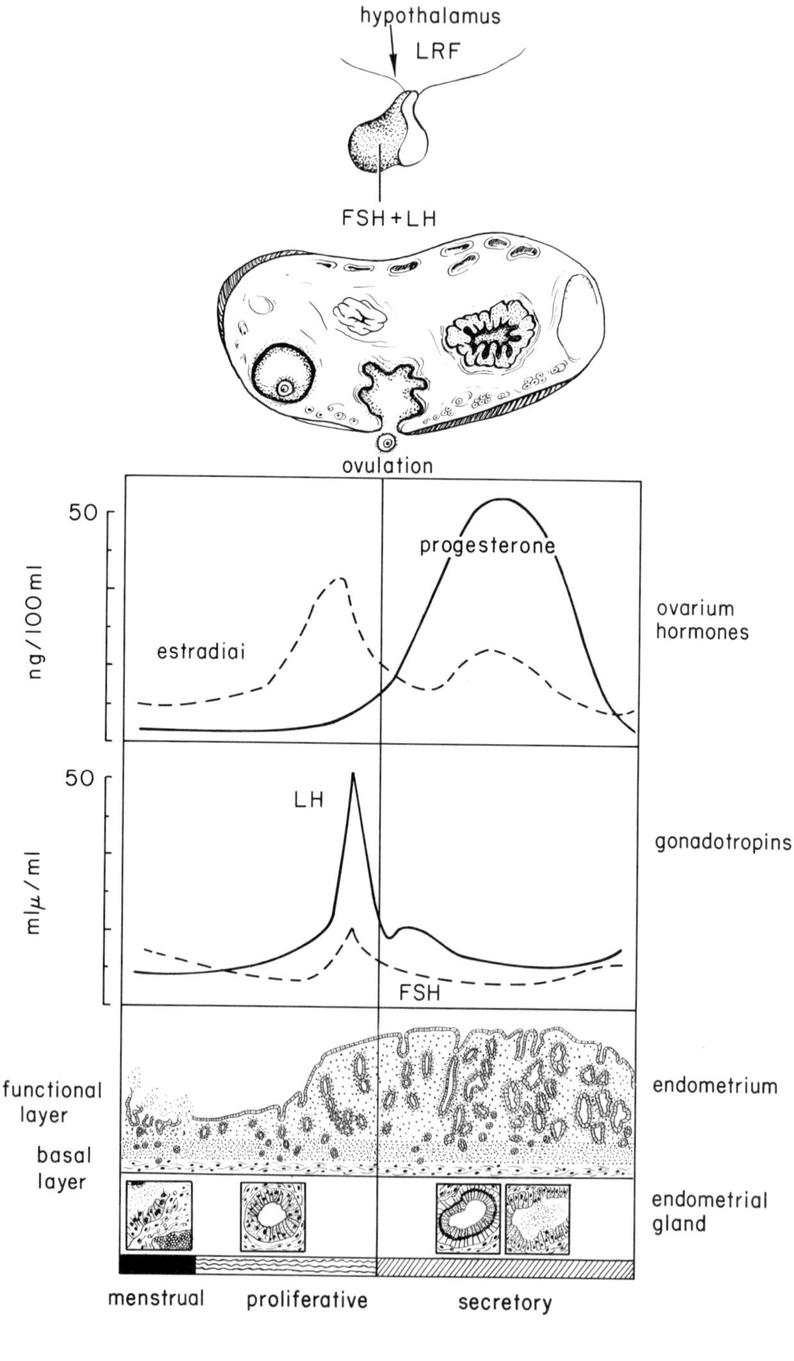
hypothalamus
LRF
FSH+LH
ovulation
50
ng/100ml
progesterone
estradiai
ovarium hormones
50
mIµ/ml
LH
gonadotropins
FSH
functional layer
basal layer
endometrium
endometrial gland
menstrual
proliferative
secretory

Figure 25. Cyclical changes in ovarian steroids and gonadotropins in peripheral plasma in relation to ovarian and endometrial changes of nonpregnant women throughout the menstrual cycle. Ovulation is presumed to take place on day 14 of an "idealized" 28-day cycle, *i.e.*, 14 days prior to the onset of subsequent menstrual bleeding. The LH shows a rise 24 hours before ovulation and a peak 16 hours prior to ovulation. The preovulatory peak of estradiol occurs some 12-24 hours before LH surge.

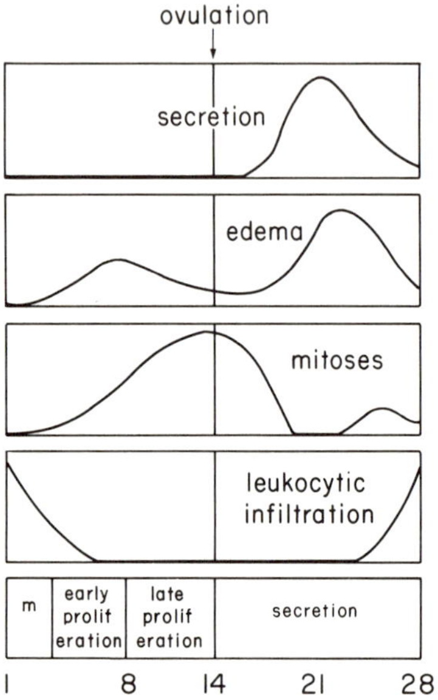
ovulation
secretion
edema
mitoses
leukocytic infiltration
m
early prolif eration
late prolif eration
secretion
1
8
14
21
28

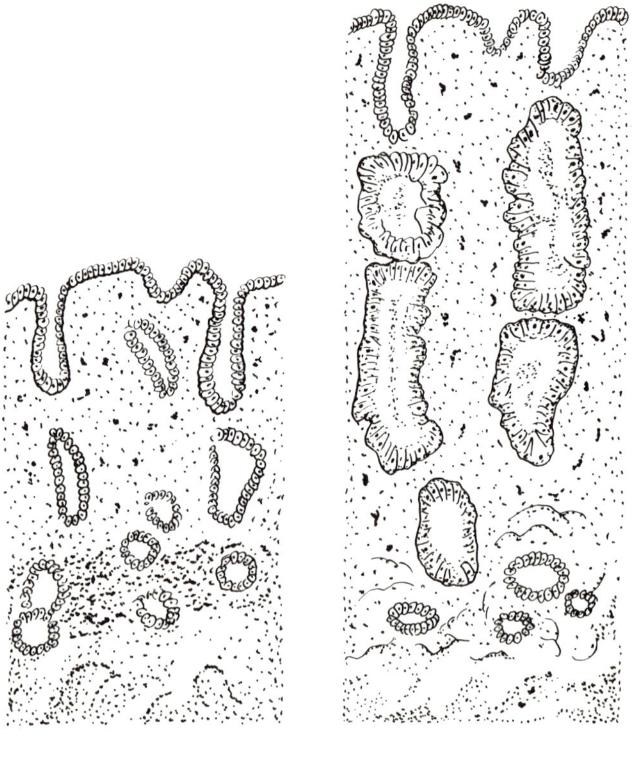

Figure 26. Some criteria for dating an endometrial biopsy during the menstrual cycle. Each curve represents approximate quantitative changes in the following morphological parameters:

(a) Secretion: visible secretion in lumen of endometrial glands;

(b) Edema: edema of stroma during proliferative phase varies with the individual; edema which accompanies secretion is more constant;

(c) Mitosis: mitosis in endometrial glands and in stroma indicate proliferation;

(d) Leukocytic infiltration: polymorphonuclear leukocytes infiltrate two days before the onset of menstrual flow. (Adapted from the literature of Herlig and Noyes, *Am. J. Obster. Gynecol.,* 1960.)

Table XII

Ovarian Endometrial and Hormonal Changes during Different Phases of a 28-day Menstrual Cycle

Phases	*Days*[a]	*Changes in Ovulation*	*Hormonal Changes*	*Changes in Endometrium*
Menstrual Bleeding	1-4	Corpus luteum regresses	Decline in estrogen and progesterone secretion	Spiral arteries constrict; endometrium breaks down
Early Proliferative	5-7	10-20 new follicles develop; most follicles degenerate, a few grow further	Developing follicles secrete estrogen; progesterone remains low throughout this phase	Endometrium is thin; repair of endometrial tissue begins; endometrial glands are straight; new arteries grow into endometrium
Late Proliferative (follicular)	8-13	One follicle matures	Developing follicles secrete more estrogen; estrogen surge occurs just before ovulation	Endometrial growth; spiral arteries grow further into proliferative endometrium; endometrial glands grow and become tortuous
Ovulation (midcycle)	14	Stigma develops; final maturation and rupture of follicle	Decrease in estrogen	
Secretory (luteal) (progestational)	15-25	Corpus luteum develops	Increase in progesterone; second surge in estrogen, but less than before	Endometrium thick, velvety and pale; endometrial glands filled with secretion; active secretion of endometrial fluid
Late Secretory	26-28	In absence of a fertilized egg corpus luteum regresses, new follicles develop	Progesterone still high, but declining; estrogen levels low	Spiral arteries constrict; involution of endometrium

[a] Day 1 represents the first day of menstrual bleeding.

the endometrial glands become relatively scarce. In the absence of a fertilized ovum in the uterus, the corpus luteum and endometrium regress, and menstruation resumes about 14 days after ovulation.

Myometrium

During nonpregnancy, the myometrium undergoes an active and quiescent phase. The active phase at the time of menstruation is compatible with the expulsion of blood and endometrial debris. During the quiescent phase following ovulation and fertilization, conditions are favorable for possible implantation of the blastocyst.

Cervix

There are cyclical changes in the diameter of the external os; dimensions of the cervical canal; tissue vascularity; and quantity and biophysical properties of cervical mucus. During the proliferative stage of the cycle, there is a progressive increase in cervical vascularity, congestion and edema, as well as secretion of cervical mucus. All these changes reach a peak at ovulation, creating optimal conditions for sperm transport in the female reproductive tract. The external os reaches a diameter of 3 mm at the time of ovulation, then declines to a minimum of 1 mm.

ENDOCRINE CONTROL OF MENSTRUAL CYCLE

The menstrual cycle is regulated by ovarian estrogen and progesterone, which are under the influence of interaction of gonadotropins of the anterior pituitary and the gonadotropin-releasing hormones of the hypothalamus (Figure 27). The sequence of events in hormonal regulation of the menstrual cycle is summarized as follows:

1. The hypothalamus has localized centers which regulate the secretion of gonadotropin FSH and LH from the anterior pituitary gland which, in turn, controls the maturation of ovarian follicles.

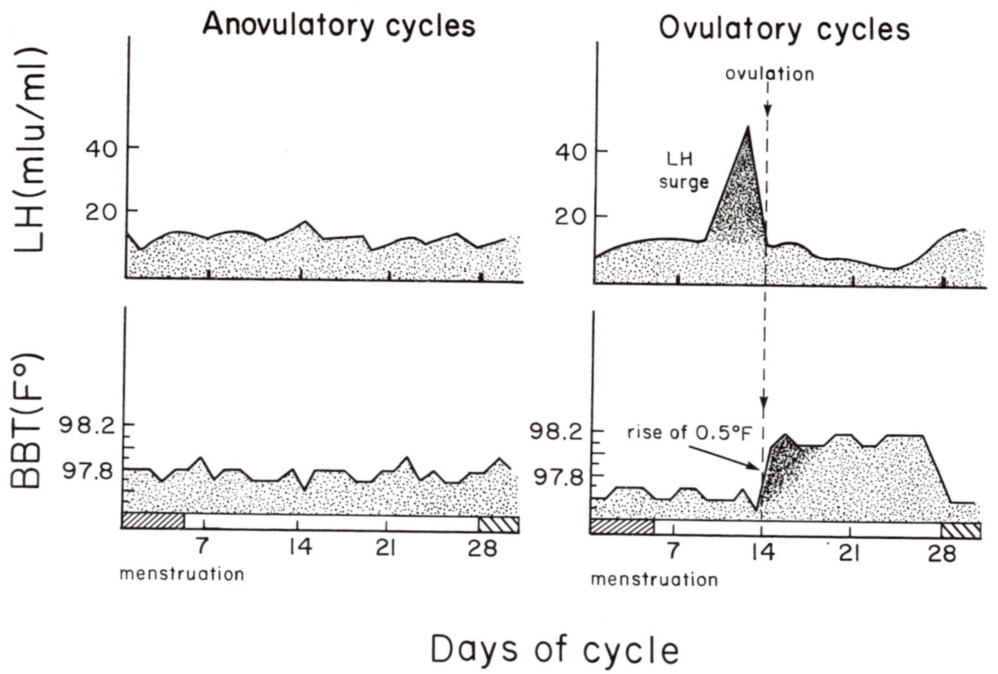

Figure 27. Ovulatory menstrual cycles are characterized by LH surge at midcycle and a concomitant rise in basal body temperature (BBT). As a result of thermogenic effect of progesterone, midcycle LH surge and BBT rise are lacking in anovulatory cycle. BBT is biphasic in most ovulatory and monophasic in anovulatory cycles.

2. At the onset of menstrual bleeding, a decline in the levels of estrogen, by feedback, causes a rise in FSH levels, characteristic of the early follicular phase.
3. The maturing ovarian follicles produce increasing quantities of estrogens.
4. As the follicles mature, rapid increase in estrogens causes a sudden increase of LH, which triggers ovulation. A remarkable surge of LH at midcycle precedes ovulation by some 16 hours.
5. Following ovulation, estradiol and progesterone (secreted by new corpus luteum) rise to reach a peak at approximately the middle of the luteal phase and then decline, gradually at first, then rapidly (Figure 25).
6. During this period, serum gonadotropin levels reach their lowest point of the entire cycle, as a result of negative feedback.
7. A premenstrual decline in luteal function reduces steroid production.
8. Before menstrual bleeding, FSH rises, and a subsequent cycle is initiated.

Menstruation can be prevented by estrogens and progesterone injections, and menstrual flow resumes within two days after withdrawal of progesterone and within four days after withdrawal of estrogen. This phenomenon is referred to as "withdrawal bleeding." Thus, ovarectomy precipitates menstrual flow within one to two days. The administration of estrogen or progesterone could prevent or stop postcastration menstruation.

CYCLICAL CHANGES IN THE OVARY

Follicular Growth

Follicular growth may be divided into three stages according to the appearance, size, rate of growth and other histologic characteristics, namely, primary, secondary and tertiary follicles (Figure 28).

During the preovulatory phase, some 3 to 30 primary follicles in each ovary begin to mature under the influence of FSH. The granulosa cells surrounding the egg increase in number

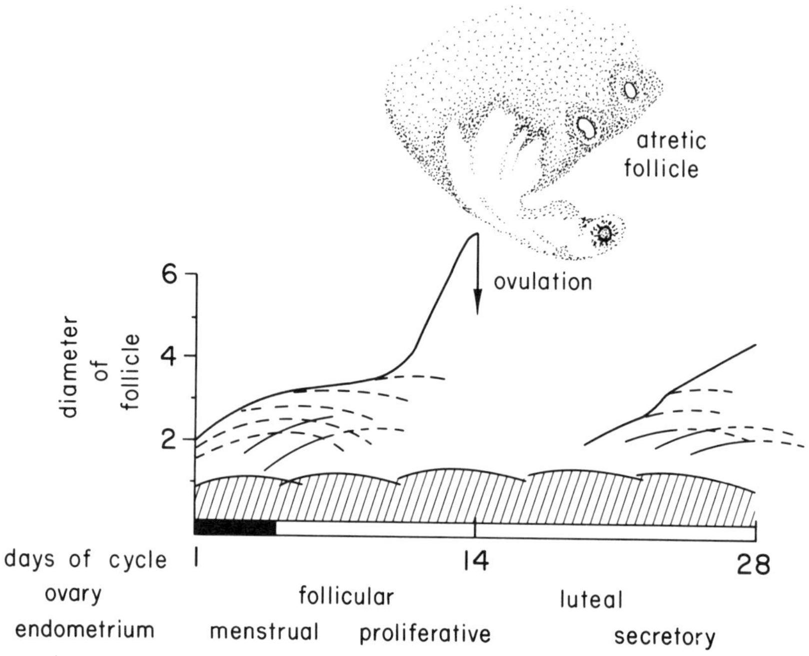

Figure 28. Diagrammatic illustration of the follicular cycle in relation to ovulation. The heavy solid line represents the diameter of the largest follicle. Other solid and broken lines represent concomitant growth and atresia of follicles that do not ovulate.

through mitotic activity and become multilayered, and the adjacent connective tissue cells differentiate into the theca interna. Some 20 follicles complete primary and secondary stages of growth during the 8 to 9 days following menstrual bleeding. The follicles proceed further, but only one eventually ovulates, while the other follicles degenerate. During the last few days of the follicular phase, most of the developing follicles undergo atresia, while the dominant Graafian follicle rapidly expands to reach an ovulatory size. Most of the estrogens secreted during these last few days are derived from this follicle. Ovulatory cycles are characterized by an LH surge at midcycle and a concomitant rise in basal body temperature as a result of the thermogenic effect of progesterone (Figure 27).

As the ovary ages, the number of follicles diminishes and the number of corpora albicantia increases, while the stroma gradually becomes fibrous.

Follicular Atresia

During fetal life and from birth until sexual maturity, waves of ovarian follicles grow and become atretic without reaching ovulatory size. Each successive wave of follicular growth reaches a greater degree of development until, finally, at the time of menarche, endocrine conditions are optimal for the occurrence of the first ovulation. After menarche, not all ovarian oocytes develop to the mature stage; for every egg that matures and ovulates, several follicles start to develop, but never reach maturity. Follicular atresia continues throughout the reproductive life until the follicles and oocytes disappear from the ovary during menopause. Follicular atresia is a continuous process throughout the menstrual cycle (Figure 28). It seems that the constant growth and atresia of follicles play a physiological role in the reproductive cycle other than to deplete the ovary of its oocytes.

CHAPTER 7

OVULATION

It is estimated that both ovaries at birth contain some one million oocytes. After birth, the number of oocytes decreases until the age of menopause (Figure 29). During the reproductive life, some 300 oocytes reach ovulatory stages (Table XIII). Most women ovulate about day 14 of the cycle, but there are cases in which ovulation occurs very early or very late in the cycle.

MORPHOLOGY OF OVULATION

Ovulation is a dynamic phenomenon which requires disruption of the vascular system and displacement of the connective tissue in the wall of the Graafian follicle. A few hours before rupture, the ovulatory follicle reaches a maximal diameter of 10-20 mm. As the enlarging follicle begins to bulge from the surface of the ovary, the vascularity of the follicular surface increases, except in its center, which seems devoid of blood vessels.

The inner layers of the follicular wall protrude through a gap to form a papilla or thin translucent stigma. The stigma thins out, bulges on the surface of the ovary, and becomes completely avascular. At ovulation, the bulging stigma ruptures at the apex, releasing some of the follicular fluid. The egg is released while embedded in a solid mass of follicular cells–the "cumulus oophorus" which protrudes into the fluid-filled antrum (Figure 29). The cumulus oophorus is made up of several layers of cells randomly dispersed in a gelatinous matrix and contains

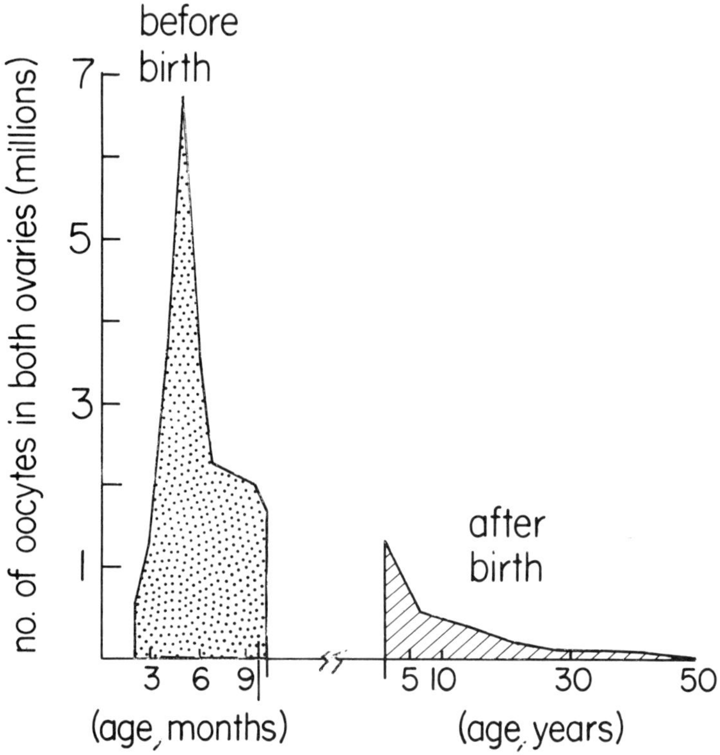

Figure 29. The number of ovarian oocytes in both ovaries before and after birth. Note the remarkable increase in the number of oocytes in the fetal ovaries until the sixth month of pregnancy, followed by a significant decline. After birth the number of oocytes decreases further until the age of menopause. There are no new oocytes formed after birth. (Data from Block, E. *Acta Anat.* 14:108, 1952; and T. G. Baker, *Proc. Roy. Soc.* 158:417, 1963.)

the egg surrounded with a layer of corona radiata cells (Figure 30). The ovulation point appears as a centrally located crater-like stigma containing follicular cells, erythrocytes and mucin-like substance. The rupture of the follicle is not necessarily associated with the release of the egg.

The ovary, attached to the back of the broad ligament, lies free in the peritoneal cavity. The oviduct curls over the ovary to facilitate egg pick-up by the fimbriae. At the time of ovulation,

Table XIII

Estimated Potential of the Human Ovary

Parameter	*Approximate Value*
Number of oocytes in both fetal ovaries	7,000,000
Number of oocytes in both ovaries at birth	1,000,000
Number of oocytes that undergo some development during reproduction	8,000
Number of menstrual cycles during reproductive life of women	400
Number of ova shed during reproductive life of women	300
Number of ovulatory cycles that could lead, in practical terms, to normal viable pregnancies	25
Number of primary follicles that start to develop during each menstrual cycle	3-30
Number of ova shed at time of ovulation	1 (rarely 2)

the ovum, together with its surrounding cells, is released into the peritoneal cavity, and swept into the ostium of the oviduct by the action of the motile kinocilia of the fimbriae. Immediately after ovulation, the ruptured follicular cavity becomes filled with lymph and blood from broken thecal vessels.

MECHANISMS OF OVULATION

Growth, maturation, ovulation and luteinization of the Graafian follicle depend on appropriate patterns of secretion, sufficient concentrations and appropriate ratios of FSH and LH in the serum. Under the influence of FSH, LH and possibly estrogen, ovarian follicles reach maturity and secrete estrogens, which act upon the hypothalamus to initiate a preovulatory LH surge. Some 16 hours before ovulation, there is a dramatic increase in the level of circulating LH (X10) and to a lesser extent, FSH (X2). The number of ovarian follicles that develop and the proportion that subsequently degenerates are regulated by the

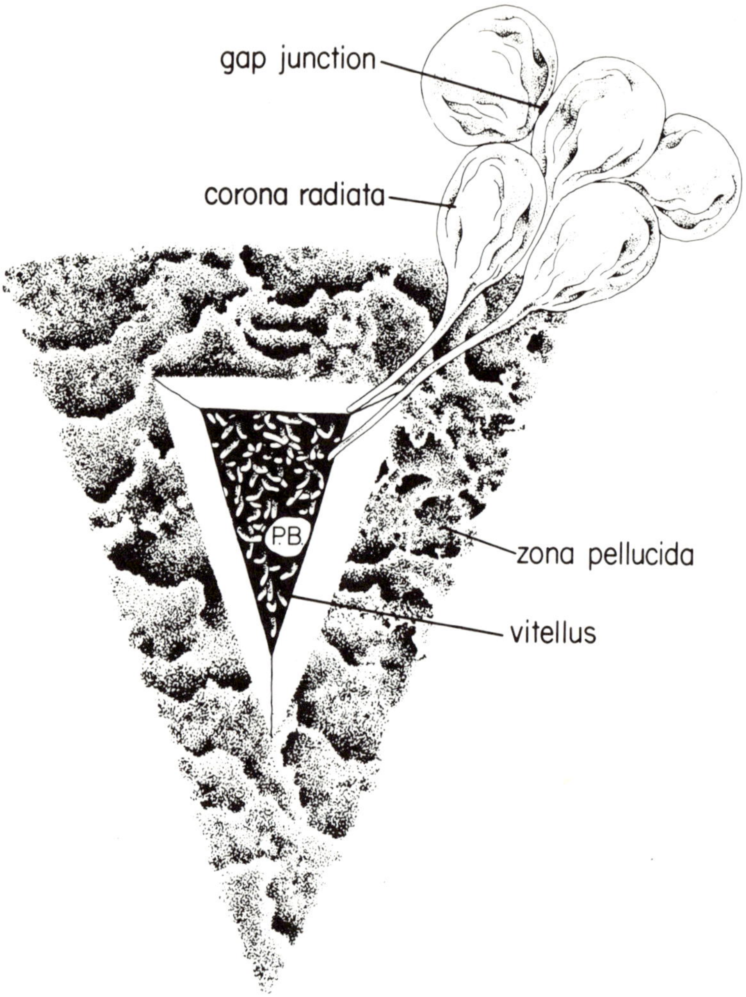

Figure 30. Diagrammatic illustration showing the corona radiata cells extending cytoplasmic processes through the zona pellucida to connect the vitellus of the egg, which is covered with abundant microvilli. The polar body (PB) is situated in the perivitelline space.

relative proportion of FSH and LH. The nature of the signal responsible for the initial development of an immature follicle is unknown.

Gonadotropins provoke a sequence of biochemical and physiological events which culminate in ovulation. LH binds to receptors in the follicle and stimulates local synthesis of prostaglandins. Prostaglandins stimulate ovarian contractility and ovulation, whereas inhibition of prostaglandin synthesis with indomethacin prevents gonadotropin-induced ovulation. Ovulation has been inhibited by agents which inhibit ovarian contractions such as adrenergic blocking agents, indomethacin and antihistamine.

Waves of contractions occur in the ovary during the follicular phase. Before ovulation, there is an increased contractility of ovarian smooth muscle and an increase in prostaglandin content of the ovarian follicle. LH stimulates the synthesis of steroids by the ovarian follicle. It is possible that these steroids participate directly in the ovulatory process by stimulating the formation of enzymes which degrade the follicular wall.

The ovary is richly supplied with adrenergic and cholinergic innervation particularly in the theca externa of the follicle. The follicular wall contains an extensive system of smooth muscle cells, which are innervated by autonomic nerves. Activation of these nerves can modify the tension of the follicle wall and affect intrafollicular pressure.

Several physiological, biophysical and biochemical mechanisms facilitate the release of the ovum, *e.g.*, contraction of ameboid evagination of granulosa cells and corona radiata, presence of smooth muscle tissue and autonomic nerves in the ovaries, contraction of fibromuscular cells of perifollicular region, peristaltic contraction of oviductal musculature and of uteroovarian ligaments, and fluid currents and countercurrents created by kinocilia of oviductal fimbriae (Figure 31).

OVULATION DETECTION AND PREDICTION

Ovulation detection is an important gynecological examination usually performed in Infertility Clinics. The presence of a mature ovarian follicle is generally not detectable by rectovaginal

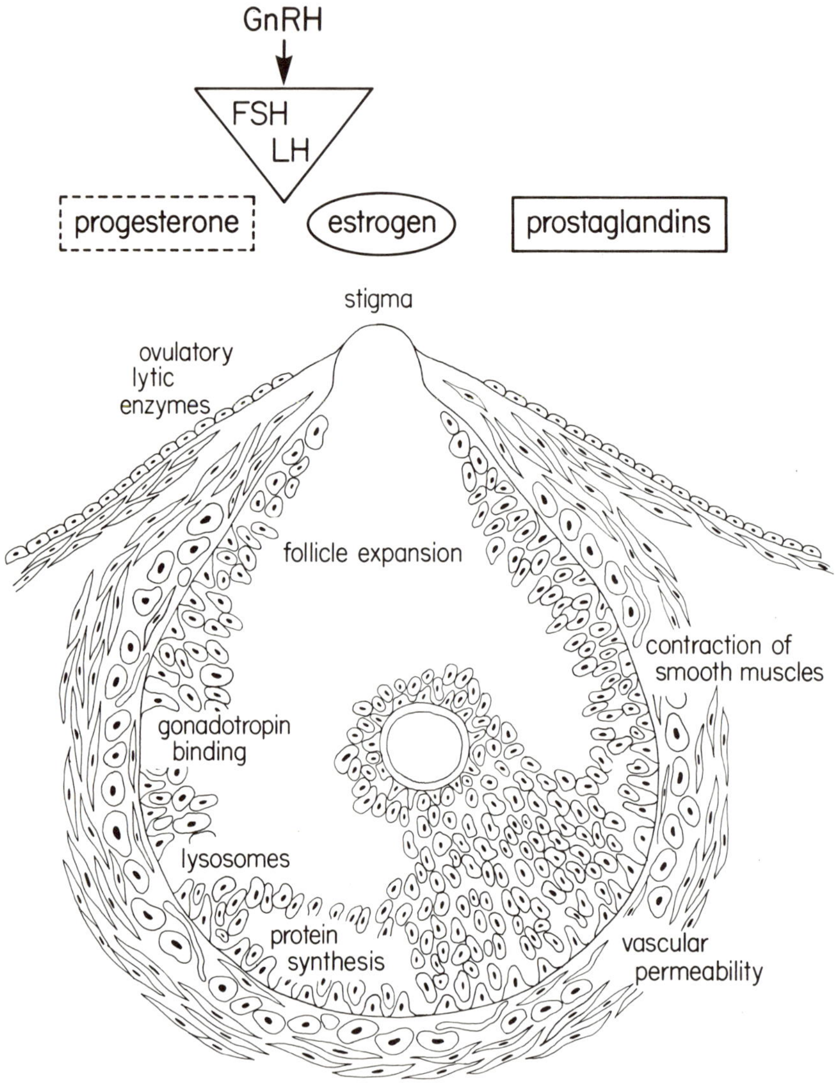

Figure 31. Diagrammatic illustration of the various physiological and biochemical intrafollicular and extrafollicular mechanisms of ovulation as affected by hormonal milieu.

examination. A sudden and brief pain during midcycle in one fossa may be a sign of ovulation.

The most accurate method to detect ovulation is the incidence of pregnancy or direct visualization of the ovarian surface with subsequent recovery of an egg from the female reproductive tract. Several indirect clinical and laboratory techniques have been used to predict and/or detect ovulation: biophysical and biochemical changes in cervical mucus, basal body temperature shift (BBT), appearance of secretory endometrium, preovulatory endocrine profile, ultrasonic examination of the ovaries and laparoscopy. The laparoscopic examination of the ovary facilitates a direct visualization of the ovarian surface and inspection of the corpus luteum. Definite diagnosis of ovulation is still very difficult by indirect laboratory techniques, particularly in women with pelvic endometriosis.

Midcycle Cervical Mucus

The weight of cervical mucus increases gradually during the follicular phase, reaches a peak one day after the midcycle estradiol peak and decreases after ovulation. At ovulation, the cervical epithelium is under maximal estrogenic influence: the cervical mucus is most copious and its water content approaches 98%. A simple technique of serial evaluation of the cervical score can be used to detect ovulation. The cervix is inspected with a vaginal speculum, and a specimen of mucus is taken with a glass pipette. A cervical score (0 to 3) is made for each of four parameters: (1) amount of mucus, (2) its Spinnbarkeit, (3) its ferning, and (4) the degree of opening of the external cervical os.

Cervical mucus is made up of two phases of distinct biophysical properties. The "gel" phase, an interwoven network of glycoprotein (mucin), comprises microfibrils and macrofibrils of variable diameter. The "sol" phase, aqueous cervical plasma, is found both as a fine microlayer adjacent to the epithelium of cervical crypts in which kinocilia beat, and within the micellar cavities in which sperm are transported by their directional motility to the uterine lumen.

During the menstrual cycle, cervical mucus undergoes several biochemical cyclical changes that are correlated closely with ovulation, *e.g.*, concentration of water, albumins, globulins, transferrin, lactoferrin, enzymes (alkaline phosphatase and amylase), glucose, maltose, glycogen, free fatty acids, mono-, di-, and triglycerides, phospholipids and electrolytes (Na, Ca, K, Mg, Cu, Zn, and Fe). Under the influence of midcycle estrogens, the concentration of most of these compounds declines, with the exception of water and certain lipids, which increase. The concentration of sialic acid and the activity of a few enzymes decrease at midcycle before LH surge and begin to rise after ovulation. It is possible that self-detection of certain enzymes in cervical mucus, particularly that of alkaline phosphatase, may provide a practical method of ovulation prediction.

Basal Body Temperature (BBT)

BBT is used for ovulation detection to aid both fertility and infertility problems. The cyclic thermogenesis in the female is a central phenomenon and is affected by emotion, physical activity and ovarian hormones. A rise in temperature is associated with ovulation and an LH surge (Figure 32).

Monophasic BBT refers to curves without a premenstrual elevated temperature but with irregular daily temperature fluctuation of 0.1 to 0.2°C. Monophasic BBT is common in the first 20 cycles after menarche, and during premenopause as a result of ovarian dysfunction, short life span of corpus luteum or deficient luteal phase. During the reproductive life, monophasic cycles appear singly or in clusters interspersed with biphasic BBT. Monophasic BBT is more common in short (<17 days) and long menstrual cycles (>60 days) than in normal cycles (25 to 32 days). Inadequate techniques of measurement, charting or interpretation of BBT may interfere with diagnosis of anovulatory cycles.

In the symptothermal method, which is more effective than BBT, fluctuations in basal body temperature are combined with cyclical changes in cervical mucus, premenstrual pain and other related symptoms (Figure 33).

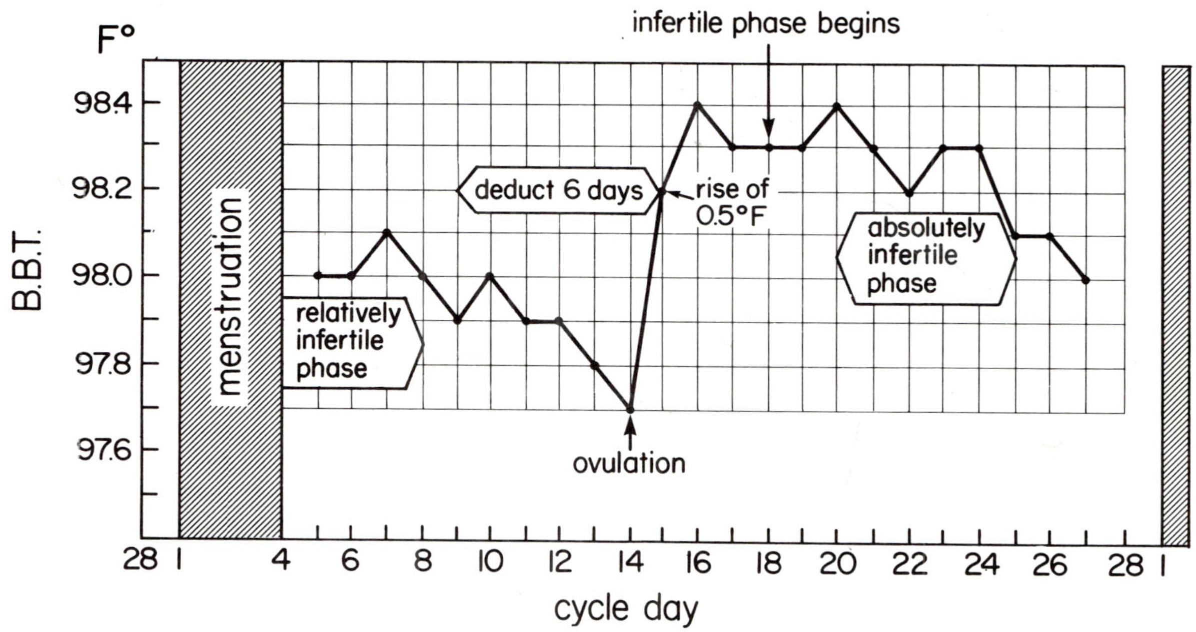

Figure 32. Cyclical changes in basal body temperature (BBT) during the menstrual cycle to determine the postmenstrual infertile phase: (a) postmenstrual and follicular phase: BBT is relatively low; (b) at ovulation: a drop of 0.2°F followed by a rise of 0.5°F. This rise is maintained until about 48 hours before menstruation, when BBT falls to its former level. The rise of temperature appears to be a progesterone effect and does not occur in anovulatory cycles; (c) natural family planning method: note infertile and fertile periods. Fertilization of the egg seems possible during the fertile period when the association of sperm and eggs is possible around the time of ovulation.

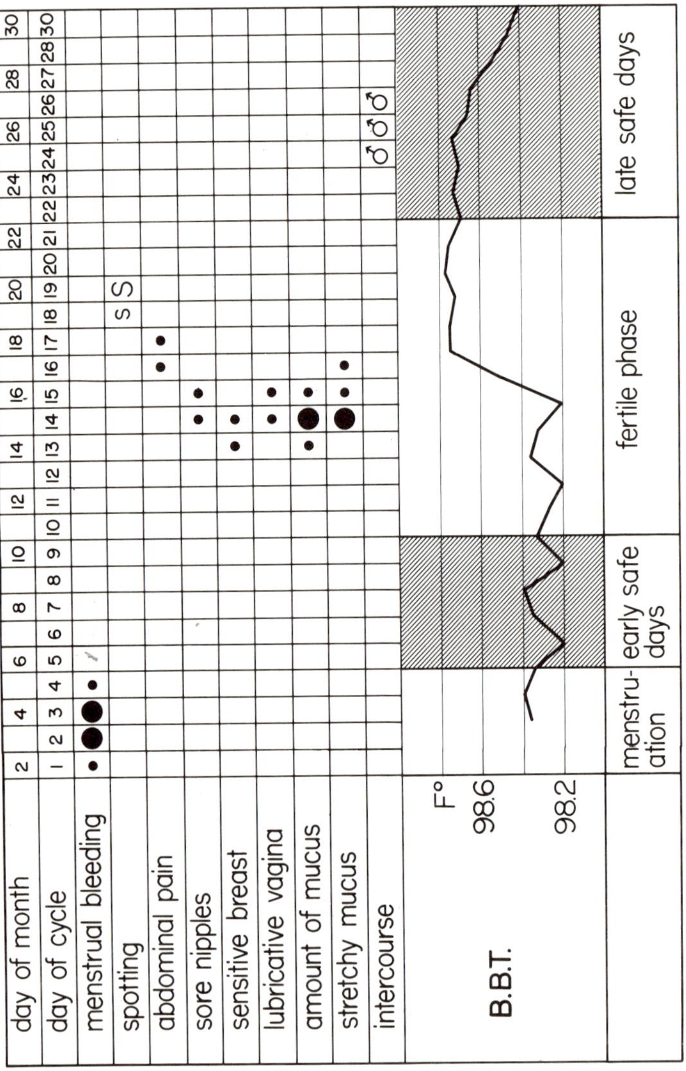

Figure 33. A typical symptothermal chart showing record of various physiological parameters and monthly events (top); and temperature chart (bottom) as used as an aid in identifying early and late infertile periods throughout the menstrual cycle.

Breast Temperature

There is a peak in breast temperature 2 to 4 days before ovulation followed by a postovulatory increase in breast temperature. Nonbreast skin temperatures do not exhibit the peak observed at the breast. Increases in breast circulation underlie the midcycle peak of breast temperature caused by preovulatory estrogen surge.

Ultrasonics of Graafian Follicles

Ultrasonic depiction of the ovaries and the preovulatory growth of a large Graafian follicle is used to estimate the optimum time of conception, especially in stimulated cycles, and to avoid a severe hyperstimulation syndrome. As the fresh corpus luteum seems to have the same ultrasonic appearance as a follicle, except for being somewhat smaller in diameter, reliable timing of ovulation is possible by this technique.

Ovulation Prediction for Artificial Insemination

Difficulty in prediction of ovulation, improper timing of inseminations and poor insemination techniques have resulted in low pregnancy rates in patients undergoing artificial insemination (AI). Ovulation may be regulated and predicted within two weeks of its occurrence allowing time to arrange for a donor and to schedule the inseminations. The woman is given 50 mg of clomiphene daily for five days. The first treatment can be initiated on the fourth, fifth, or sixth day of the menstrual cycle so that expected ovulation would occur on weekdays.

OVULATION INDUCTION

Anovulation may be caused by dysfunction of hypothalamo-pituitary-ovarian axis, prolactinemia, or minor imbalance in thyroid adrenal or pancreatic function. Ovulation may be induced by correcting associated subclinical endocrinopathies or concomitant systemic diseases, which increase the sensitivity of target organs to stimulation. The choice of a specific therapeutic measure depends on identifying an endogenous underlying factor causing anovulation. Proper classification and selection of patients, dosage, timing and regimen of ovulation induction is a prerequisite for achieving high success rates.

Ovulation can be induced by human gonadotropins, clomiphene or GnRH and its analogues. All these agents stimulate gonadal function directly or indirectly through different mechanisms and are capable of inducing ovulation in properly selected infertile patients. A proper classification of patients will help to select the appropriate type of therapy.

Proper selection of a modality of therapy for the induction of ovulation should be based on the proper interpretation of simple laboratory studies. For example, clomiphene citrate is most effective in patients with normal plasma FSH and LH levels and good cervical mucus, whereas human gonadotropins provide the best results in hypogonadotropic estrogen-deficient women.

Ovarian Hyperstimulation Syndrome

High dosages of exogenous gonadotropins may cause increased capillary permeability as a result of increased intermediate metabolites by the stimulated ovary. Ovarian hyperstimulation syndrome may be a serious complication of ovulation induction with gonadotropins. Severe cases are associated with ascites and thromboembolic phenomena.

Gonadotropin-Resistant Ovary

During premature menopause, the ovaries contain several early follicles, which do not respond to high levels of endogenous or exogenous gonadotropins. The physiological mechanisms are not known, but this syndrome can be evaluated by examining the endocrine profile of the patient.

Failure in Ovulation Induction

Endogenous and exogenous factors may be responsible for failure to induce ovulation in infertile patients. The indiscriminate use of available drugs without prior identification and treatment of associated subclinical disorders may be responsible for high failure rates. Misleading results may be reached when ovulation detection is based on available indirect methods which cannot differentiate between "true ovulation" and "retained ruptured follicle." This may explain discrepancies between ovulation rates and pregnancy rates. However, a deficient corpus luteum or possible modification of gamete transport may be responsible for failure of ovulation induction.

Ovarian Transplantation

Ovarian transplantation involves several challenges. Functional survival of the transplant must secure dual ovarian function–endocrine and generative. Transplantation failure is primarily due to difficult direct anastomosis of ovarian branching blood vessels of small diameter and frequent clotting of the vascular channels due to thrombosis as a result of surgical manipulation. To ensure the morphological and functional survival of the graft, ovarian tissue is grafted to highly vascularized sites. Each transplantation technique results in temporary devascularization (anoxia) and deprivation of neural connections and lymphatic drainage. Ovarian tissue, including oocytes, tolerates temporary anoxia.

CORPUS LUTEUM

The corpus luteum is a temporary endocrine organ which is formed following ovulation. After rupture of the follicle, the follicular walls collapse, and the outgrowth of the enfolded walls form the corpus luteum. The granulosa cells of the follicle hypertrophy and proliferate to form luteinized cells of the corpus luteum. This process of luteinization is initiated by LH. Some follicles mature, but do not rupture and no corpus luteum is formed. If this is the case, the follicular fluid is gradually reabsorbed and the follicle regresses, forming a corpus fibrosum.

Life Span

The corpus luteum, with a maximal diameter of 1 cm, has a finite life, depending on whether conception follows. In the presence of a fertilized ovum in the uterus, implantation of the blastocyst is initiated 7 days after fertilization. Implantation is associated with the excretion of a unique hormone, human chorionic gonadotropin (HCG), produced by the trophoblast of the implanting blastocyst. Detection of this hormone in the blood transforms the pattern of a regular cycle into a conception cycle. If pregnancy follows, the corpus luteum remains functional for 3 to 4 months, then gradually degenerates.

In the absence of a fertilized ovum and an implantation site, HCG, necessary for the survival of the corpus luteum, is not produced, and the corpus luteum regresses. These morphological changes are associated with decline in the secretion of steroids leading to subsequent menstruation about two weeks after ovulation. Eventually, the corpus luteum is transformed into the corpus albicans, which appears as a barely visible tortuous white scar on the ovarian surface.

Function

Progesterone secreted by the corpus luteum prevents follicular maturation and is responsible for the structural changes in the endometrium necessary for implantation. Steroids secreted by the corpus luteum produce a feedback on the hypothalamus and this, in turn, provides a continuing sequence of menstrual cycles, which is only interrupted by pregnancy and lactation.

Maximal function activity of the corpus luteum occurs by the eighth day after ovulation. Highest levels of estrogen and progesterone at this time coincide with the time of implantation of the embryo in the endometrium. The secretion of human chorionic gonadotropin (HCG) at the time of implantation maintains luteal progesterone secretion until this function is taken over by the placenta at about the sixth week of pregnancy.

CHAPTER 8

TRANSPORT AND SURVIVAL OF SPERM AND EGGS

SPERM TRANSPORT

The normal ejaculate contains some 100 million sperm, nearly all of which degenerate and are absorbed by the female reproductive tract. Only thousands of sperm enter the oviduct, and only tens reach the site of fertilization at one time. The main sites of reduction of the number of sperm are the cervix, the uterotubal junction and the isthmus of the oviduct.

During their transport to the site of fertilization, sperm are significantly diluted with luminal secretions from the cervix, uterus and oviduct, and are susceptible to changes in the pH of these luminal fluids. The pH range of normal midcycle cervical mucus (7 to 8.5) is optimal for sperm migration and survival. Acidity and excessive alkalinity (8.5) of mucus immobilize sperm whereas alkaline mucus enhances sperm motility. Follicular, oviductal, peritoneal and amniotic fluids increase the activity and velocity of sperm.

Patterns of Sperm Transport in the Female Tract

The patterns of sperm transport and the physiological mechanisms involved are summarized in Tables XIV and XV. Three stages are recognized in sperm transport within the female reproductive tract: (1) rapid, short transport; (2) colonization of cervical reservoirs; and (3) slow, prolonged release of sperm.

Table XIV

Transport and Survival of Human Sperm

Physiological Characteristics	*Values*
Sperm transport time, from their release by Sertoli cells to appearance in ejaculate	19-23 days
Time interval for sperm transport from cervix to oviduct	5 min
Propagation of mucus secretion from cervical epithelium by ciliary action	1-7 mm/min
Wave frequency of sperm tail at 32°C	14-16 heat/sec
Velocity of sperm progression in cervical mucus	0.2-3.0 mm/min
Duration of sperm motility in female reproductive tract:	
Motility of sperm in vagina	0.5 day
Motility of sperm in cervical mucus following intercourse	2-8 days
Motility of sperm in cervical mucus following artificial insemination	7 days
Motility of sperm in uterus and oviduct	2-2.5 days
Duration of sperm fertilizability in female reproductive tract	? (probably 1 day)

Rapid Transport

Immediately after deposition of semen into the vagina, sperm penetrate the micelles of the cervical mucus where some sperm are rapidly transported through the cervical canal. This phase takes 2 to 10 minutes and may be facilitated by increased contractile activity of the myometrium and mesosalpinx during precoital excitement. Some sperm reach the internal os of the cervix within 1.5 to 3 minutes after ejaculation. Immotile sperm or particles may be transported from the vagina to the oviduct within 30 minutes.

Colonization of Reservoirs

The cervical crypts and probably the uterotubal junction constitute the major "selective reservoirs" for sperm. The orientation of micelles of cervical mucus lead to their point of secretory origin: the cervical epithelium (Figure 34). These orientation lines tend to guide a majority of motile sperm to the cervical crypts. Fewer numbers of sperm are found in the mucus of the cervical canal, the endometrial glands and the endometrial fluid. Very few sperm are found in the oviductal fluid. Concentration gradients of sperm are established in the female reproductive tract within a short time interval after intercourse.

Slow Release of Sperm

After adequate sperm reservoirs are established within the reproductive tract, the sperm are released sequentially to the site of fertilization. This slow release, which involves the intrinsic motility of the sperm and contractile activity of myometrium and mesosalpinx, continues for a prolonged period after intercourse.

Transport in the Cervix

Several functions have been ascribed to the cervix and its secretion: (1) receptivity to sperm penetration at or near ovulation and inhibition of migration at other phases of the cycle; (2) sperm reservoir; (3) protection of sperm from the hostile environment of vagina and from being phagocytized, (4) provision of energy requirements for sperm; (5) filtering of defective and immotile sperm; and (6) possible sperm capacitation.

Ejaculated sperm rapidly penetrate watery cervical mucus during midcycle, aided principally by sperm motility, as well as the micro- and macrorheological properties of mucus. The rate of sperm penetration in the mucus varies throughout the menstrual cycle from 0.1 to 3 mm/min and is maximal in the preovulatory and ovulatory stages. Sperm penetrability begins approximately on the ninth day of a normal menstrual cycle, increases gradually to a peak at ovulation, persists for 1 to 3 days and is inhibited until the following midcycle (Figure 35). There are remarkable individual variations in the length of period for

Table XV

Sequence of Events and Physiological Mechanisms Associated with Sperm Transport in the Male and Female Reproductive Tracts

Location of Sperm	*Sequence of Events*	*How This is Regulated*
Male Reproductive Tract	Sperm stored in tail of epididymis, undergo maturation	Neuromuscular factors
	At ejaculation, sperm released from epididymis, are mixed with male accessory secretions	
Vagina	Semen emission in several ejaculatory pulsations	Copulatory motor activities
	Semen mixed with vaginal and cervical secretions	
Cervix	Sperm migrate through micelles of cervical mucus	Biophysical and biochemical changes in cervical mucus
	Abnormal sperm filtered (gross selection of sperm) through mucus within cervical canal	
	Cervical crypts establish "sperm reservoir" or rid themselves of excessive sperm causing remarkable reduction in sperm number	
Uterus	Sperm are separated from seminal plasma and transported to oviduct	Contraction of myometrium
	Biochemical and physiological changes associated with sperm capacitation	Metabolic changes of sperm in uterine fluid
	Acrosomal proteinase (trypsin-like enzyme) are inactivated by trypsin inhibitors from seminal plasma	

Uterotubal	Quantitative selection of sperm	Mechanical factors
Isthmus of Oviduct	Sperm numbers reduced in narrow lumen	Mechanical factors
Ampullary-Isthmic Junction (intramural portion of oviduct)	Regulation of egg transport in oviduct to ensure fertilization	Neural
	Changes in plasma membrane of sperm (acrosome reaction)	Biochemical factors
	Sperm capacitation	
Ampulla of Oviduct	Sperm motility increases in oviductal fluid to be able to penetrate corona radiata and zona pellucida of egg	Metabolic and enzymatic mechanisms
	Reduction division of sperm and egg	
	Acrosomal proteinases released	
	Selection of egg surface (receptors?) by sperm	
Fimbriae of Oviduct	Excessive sperm are lost into peritoneal cavity	Intrinsic sperm motility

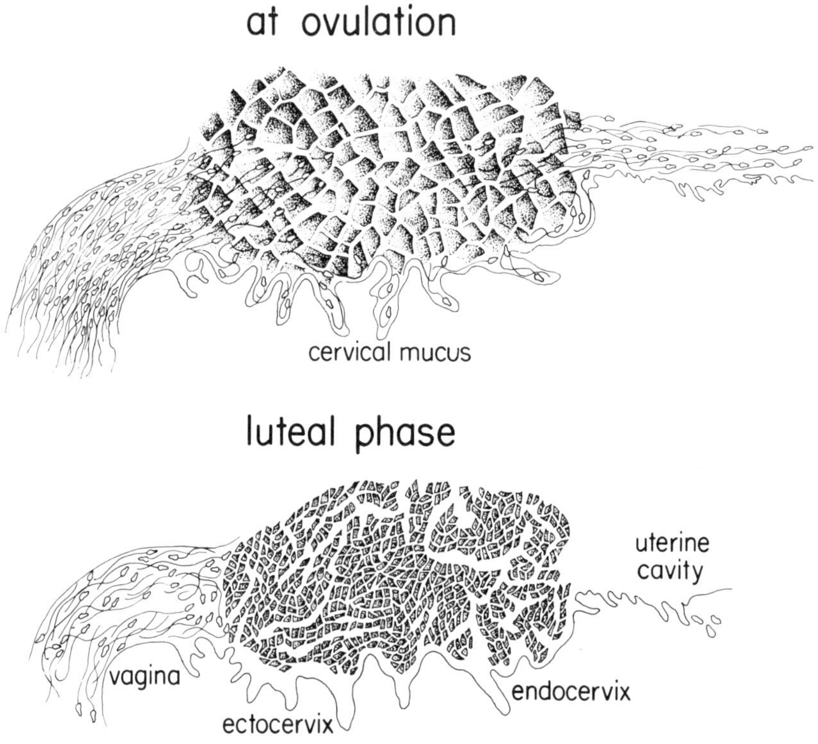

Figure 34. The cervical mucus within cervical canal makes a mesh of micelles, which permits sperm transport during ovulation and prevents sperm transport to the uterine lumen during the luteal phase.

sperm penetrability in cervical mucus. Sperm penetrability increases with the cleanliness of the cervical mucus, since cellular debris and leukocytes delay sperm penetration.

The rate of sperm penetration in the cervical mucus is also affected by the physiological integrity and density of sperm. Higher sperm density and greater motility are associated with more massive penetration into the mucus. Nonviable sperm do not penetrate cervical mucus, whereas freeze storage of sperm reduces penetrability of sperm.

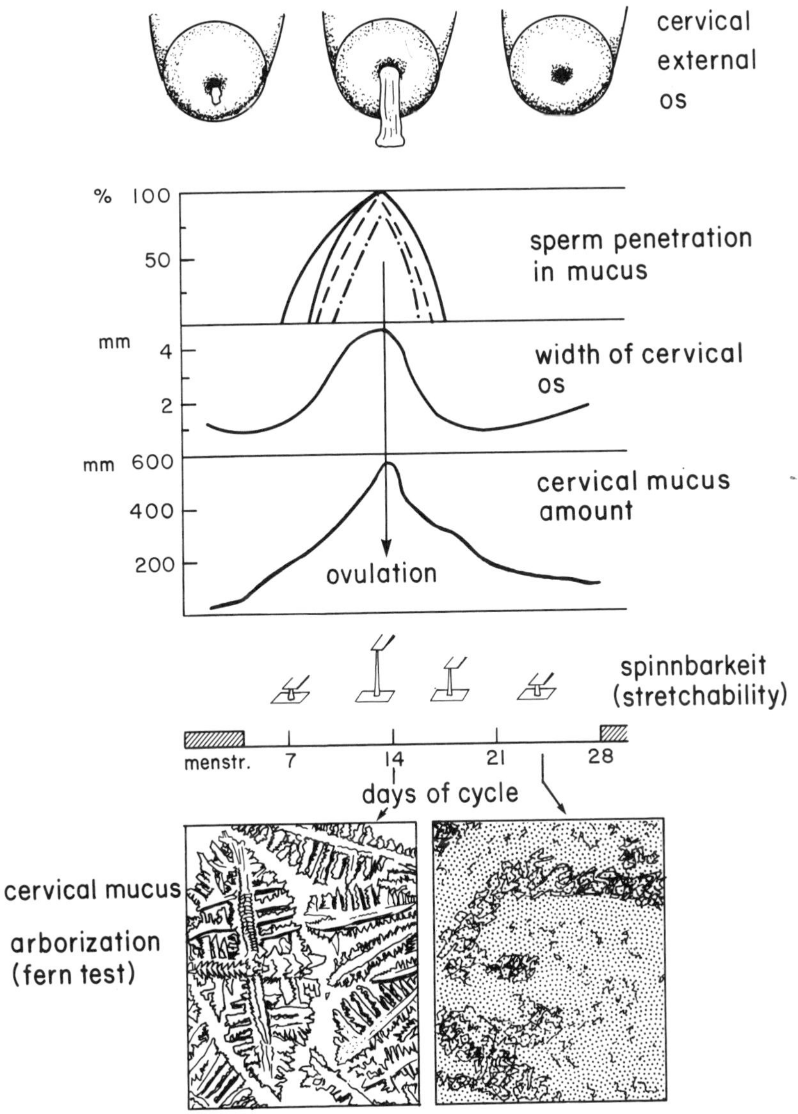

Figure 35. Cyclical changes in diameter of cervical external os, amount and biophysical characteristics of cervical mucus and penetrability of sperm in cervical mucus. During midcycle, cervical mucus shows maximal ferning upon air-drying, and maximal sperm penetration. Note individual differences in sperm penetration in cervical mucus.

Sperm are mechanically oriented toward the internal os. As the flagellum beats and vibrates, the sperm head is propelled forward into the channels of least resistance. The hydrogel of cervical mucus is made of two phases: micelles of mucin (gel), and soluble substances (sol). The aqueous spaces (sol) between the micelles allow the passage of sperm as well as the diffusion of soluble substances (Figure 36). Proteolytic enzymes seem to hydrolyze the backbone protein of some of the crosslinkages of the mucin (gel) and reduce the network to a less resistant mesh with more open channels for the passage of sperm migration.

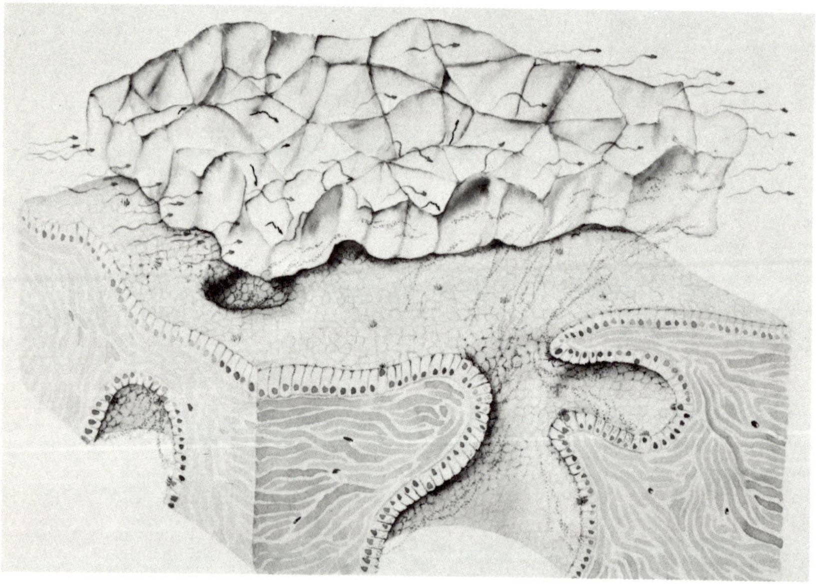

Figure 36. Diagrammatic illustration showing the production of cervical mucus. Macromolecules of glycoprotein (shown in dots) are elaborated by the secretory cells in the cervical crypts to form the mesh of cervical mucus (honeycomb structure shown on the top). A few ciliated cells are shown within the crypts and at the openings of the crypts. During midcycle, the mesh of the cervical mucus is quite wide to allow the transport of sperm from the site of ejaculation to the uterine lumen. It is possible that the micelles of the cervical mucus filters the ejaculate when nonviable sperm (shown in dark color) adhere to the mucus.

The cervical mucus secreted at the time of ovulation provides an environment suited to the maintenance of sperm metabolism. This mucus undergoes biochemical changes, such as decrease in albumin, alkaline phosphatase, peptidase, antitrypsin, esterase and sialic acid; and an increase in mucins and sodium chloride. These changes seem to make mucus suitable as a "culture medium" for sperm.

Transport in the Uterus

Maximal numbers of motile sperm are found in the uterus when intercourse occurs at or near ovulation. The contractile activity of the myometrium plays a major role in sperm transport into and through the uterus. The number of sperm in the uterine lumen is highly correlated with that in the oviducts and pouch of Douglas. In the presence of a corpus luteum, the number of motile sperm in the uterus decreases. Once the sperm reaches the uterine lumen, their further progression to the oviduct is not influenced by the stage of the menstrual cycle.

Some sperm invade the endometrial glands, whereas phagocytosis of living and probably dead sperm by leukocytes occurs in the uterine lumen.

Transport in the Oviduct

The oviduct has the unique function of conveying sperm and eggs in opposite directions, almost simultaneously. Sperm reach the oviducts within five minutes of being placed on the cervix. The ascent of sperm from the vagina to the oviduct thus occurs more rapidly than the intrinsic motility of the sperm could account for.

Sperm transport through the oviduct is controlled by several mechanisms, such as segmental peristaltic and antiperistaltic contraction of oviductal musculature; complex contractions of the oviductal mucosal folds and of the mesosalpinx; fluid movements created by ciliary actions and possibly the opening and closing of the intramural portion of the oviduct. Oviductal contractions alter the configuration of the oviductal compartments momentarily, so that fluids and sperm may be transported toward the fimbriae from one compartment to the next.

It seems that the main function of the isthmus of the oviduct is to restrain migration of excessive numbers of sperm to the ampulla and thus reduce the chances of polyspermic fertilization which is often associated with subsequent embryonic mortality.

Endocrine Control of Sperm Transport

Ovarian hormones affect the structure, ultrastructure and secretory activity of the cervical, uterine and oviductal epithelium; the contractile activity of the uterooviductal musculature; and the quantitative and qualitative characteristics of cervical mucus, and uterine and oviductal secretions. Increasing levels of endogenous estrogen during the preovulatory phase of the cycle or the administration of synthetic estrogens produce copious amounts of thin, watery, alkaline cervical mucus. Endogenous progesterone during the luteal phase of the cycle or during pregnancy produces scanty, viscous cellular cervical mucus with low Spinnbarkeit. It is possible that the cyclic changes in cervical mucus are mechanisms for protecting the female from unnecessary exposure to foreign proteins of the semen.

Sperm transport in the female genital tract is also controlled by oxytocin and the sympathetic and parasympathetic nervous system. During intercourse, the posterior pituitary releases oxytocin which stimulates the contraction of the uterooviductal musculature. Epinephrine, acetocholine, histamine and various vasoconstrictors can alter uterine contraction. Psychological stress and psychosomatic factors may inhibit sperm transport, due to the release of adrenalin, which reduces the contractility of the myometrium in response to oxytocin.

The absorption of prostaglandins from the seminal plasma into the vaginal walls may be important for sperm transport in the immediate postcoital phase. The sensitivity of the uterus to prostaglandins reaches a peak between the late proliferative and early secretory phases of the menstrual cycle.

MOTILITY OF SPERM

The normal human sperm is 50 μ long. The sperm head, which is oval-shaped and flattened in one dimension, measures 4.5 μ in length, 2.5 μ in width and 1.5 μ in thickness and contains a large nucleus. The flagellation of the tail and the hydroplane shape of the sperm head facilitate its migration through the micelles of cervical mucus.

Under normal physiological conditions and in the fully activated state, sperm show directional progressive movement, combined with a rotation along their longitudinal axis. The rotation is due to hydrodynamic effects arising from the fact that normal tail movements consist of a three-dimensional wave. There are also other types of sperm motility: (1) progressive movements with small lateral head displacements; (2) progressive movement with some irregular changes of direction, and large lateral head displacements; and (3) yawing, circular and helical motility.

Highly active sperm may beat synchronously in phase, in suspension where sperm concentration is dense. Synchrony of flagella beat may be related to fertility and may determine the optimal and minimal numbers of sperm necessary for fertilization.

SURVIVAL OF SPERM

Survival of sperm in the female reproductive tract is an important parameter in fertilization and conception. Much information is known about the duration of sperm motility, but little is known about the duration of fertilizing capacity, which is lost before motility. Sperm can remain viable for months in the uterus of hibernating mammals such as bats, and for years in some insects and reptiles. It is believed that human sperm are capable of fertilization for 24 to 48 hours after intercourse; should ovulation occur during this time, fertilization is likely. However, motile sperm have been recovered from the human uterus and oviduct several days after intercourse.

Sperm survival in the female reproductive tract is affected by several factors, *e.g.*, quality of semen, frequency of ejaculation by the male, and age and health of the male. On ejaculation, sperm are rapidly separated from the seminal plasma and are resuspended in the female luminal fluids. Thus, sperm survival differs in different regions of the female tract, the physiochemical properties of which are dependent upon the endocrine state of the female.

The acidic milieu (pH 3 to 4) limits survival of sperm in the vagina. Motile sperm persist in the cervix longer than in other regions of the genital tract. The protective action of cervical mucus ensures a constant release of motile sperm into the upper tract, following the initial rapid passage, thus increasing the chance of fertilization. Prolonged precoital stimulation of the female stimulates the production of vaginal fluid which, in turn, when mixed with cervical mucus, alters the pH of the posterior fornix of the vagina and prolongs the survival of ejaculated sperm.

LOSS OF SPERM

Although millions of sperm are deposited into the reproductive tract of the female, few ever reach the egg at the site of fertilization. Most sperm perish at the selective barriers; the vagina, cervix, uterotubal junction and oviductal isthmus. Sperm reaching the fimbriae of the oviduct are released into the peritoneal cavity, and may remain motile in the peritoneal fluid for 24 hours after intercourse.

The presence of semen in the uterine cavity initiates the leukocytic response; the appearance of polymorphonuclear leukocytes. Active phagocytosis is not noted until 8 to 16 hours after sperm deposition. Phagocytic removal of sperm from the female genital tract and the intensity of leukocytic response to various stimuli depend on the endocrine state of the female. Phagocytosis is less efficient in the progesterone-dominated phase, and intrauterine devices stimulate the presence of leukocytes in the uterine cavity with subsequent increased phagocytosis.

Damaged sperm are carried back passively through the ectocervix with the help of ciliated cells beating toward the vagina. Such sperm, advancing only a short distance into the cervical mucus column, do not reach the cervical crypts. Thus, the percentage of immotile sperm is higher in the cervical mucus than in other segments of the female reproductive tract.

TRANSPORT OF EGGS

The oviduct takes an active part in maintaining eggs and preparing them for fertilization, cleavage and early blastulation.

Pick-Up of Egg

At ovulation, the fimbriae of the oviduct comes in close contact with the ovary (Figure 37). The fimbriae move to and fro over the ovulating follicles to ensure pick-up of the egg. As it is released with its granulosa cells, the egg continues to adhere to the ruptured follicle unless swept off the ovarian surface by the fimbriae. The pick-up of the egg is facilitated by the large surface area of the fimbriae, which is lined up by beating kinocilia and by the vigorous contraction of the mesovarium and mesosalpinx. Currents of fluids caused by cilia and direct cilial mechanisms are involved. Pick-up of the egg can also occur from the cul-de-sac in some cases, as women with an ovary on one side and an oviduct on the other side sometimes become pregnant.

Transport of Egg in the Oviduct

Three stages are recognized in the transport of egg from the infundibulum of the oviduct to the uterus:

1. rapid transport through the ampulla;
2. retention for 1 to 2 days at the ampullary-isthmic junction where fertilization occurs (tube locking); and
3. fairly rapid transportof the egg through the isthmus and uterotubal junction.

After its retention at the ampullary-isthmic junction, the egg passes fairly rapidly through the isthmus, aided by peristaltic contractions of the circular muscle innervated by adrenergic nerve terminals. The egg is retained in the final segment of the isthmus until the uterotubal junction relaxes, and the egg is transported to the uterus through peristalsis and directional flow of the oviductal fluid (Figure 38).

The entry of the egg into the uterus is synchronized with the beginning of the luteal phase, which occurs three days after ovulation. It is critical that fertilized eggs reach the uterus at an

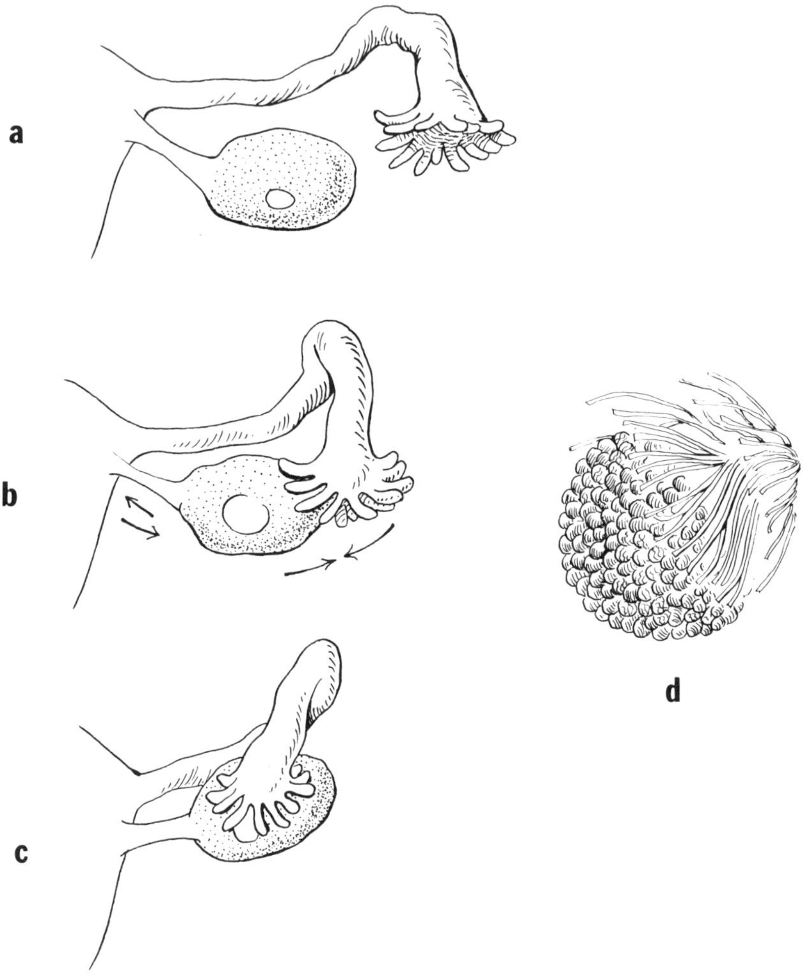

Figure 37. Diagram showing contraction of fimbriae in relation to ovarian surface to facilitate egg pick-up by the infundibulum (a,b,c). Note the contractions of the uteroovarian ligament and the proximation of the fimbriae to the ovarian surface. The rough surface of the egg surrounded by granulosa cells is picked up with the aid of beating ciliated cells of the fimbriae (d).

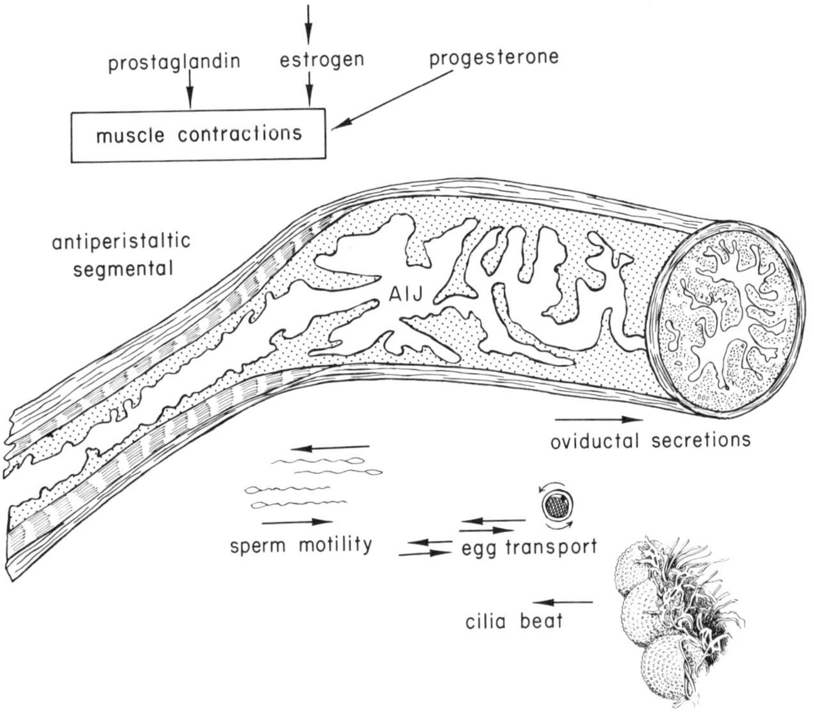

Figure 38. Diagram showing forces in oviduct which may be involved in gamete transport. The arrows in the ampulla (wide lumen) represent ciliary beat toward the ampullary isthmic junction (AIJ) and the flow of oviductal fluid toward the fimbriae. The arrows in the isthmus (narrow lumen) represent flow of oviductal fluid from uterotubal junction (UTJ). Note differential effect of estrogen and progesterone on forces in ampulla and isthmus.

appropriate progestational stage of the menstrual cycle in order to implant in the uterus. Both estrogen and progesterone must be present at appropriate levels to ensure the normal rate of egg transport.

The delay in ovum transit is an integral part of the sequence of events leading to successful fertilization and implantation. The duration of the pause of the egg appears to be rather precisely timed, and successful implantation depends upon this timing within fairly narrow limits.

Regulation of Egg Transport

Muscular contractions of the muscular wall play a major role in egg transport in the oviduct. Oviductal peristalsis causes hesitant progression of the egg, facilitates mixing of oviductal contents, promotes fertilization and helps denude the egg.

Segmental, peristaltic and antiperistaltic contractions move the egg backward and forward in a rotary fashion. Contractions of the circular musculature compress and elongate the cumulus mass. As the segmental peristaltic wave develops, the egg and surrounding cumulus mass is "milked" forward, with intermittent resting intervals; the process continues and the egg is transported to the ampullary-isthmic junction. These mechanisms can be altered by exogenous and by natural estrogens and progestins, prostaglandins and other drugs influencing the contraction of smooth musculature, such as reserpine. Adrenergic nerve fibers also control muscular activity of the oviduct.

Ovarian hormones control the activity of an electric pacemaker, which is localized around the ampullary-isthmic junction. Ovariectomy reduces the rate and amplitude of oviductal contractions. The degree of reduction in contraction depends on how recently ovariectomy has been performed.

Tube locking of the egg at the ampullary-isthmic junction may be achieved by several mechanisms, *e.g.*, temporary inactivity of the cilia, localized edema of the isthmic region, adovarian activity, constriction, or inactivity of the isthmic circular muscles.

The transport of the egg in the oviduct is also regulated by several physiological factors, creating currents and countercurrents of oviductal fluids:

1. quantitative and qualitative changes in oviductal secretions throughout the menstrual cycle;
2. beat of kinocilia in oviductal compartments of variable size and shape; and
3. constant change in the diameter of the oviductal lumen in different segments as a result of contractions of oviductal musculature and of the mesosalpinx (Figure 38).

SURVIVAL OF EGGS

The ovum sheds its first polar body while still within the preovulatory follicle. The cumulus and corona radiata cells remain around the egg until fertilization, after which it becomes dispersed, presumably by the enzymes of the sperm. Denudation of the corona may limit the fertilizable life span of the ovum. The ovum has a finite fertilization life span of approximately 12 to 24 hours. Some pregnancy wastage has been attributed to fertilization of an aged ovum.

The ovum has two distinct membranes: the vitelline membrane and the zona pellucida. The zona pellucida is a noncellular layer of mucopolysaccharide, which is deposited at the periphery of the egg while in the ovary. The zona pellucida has important functions at the time of fertilization and it protects the egg during cleavage.

It is possible to culture follicular ova through complete maturation, fertilize them *in vitro*, and culture them further so that they can be transplanted to a host uterus for subsequent development.

CHAPTER 9

FERTILIZATION AND IMPLANTATION

FERTILIZATION

Fertilization involves the union of the DNA of the nucleus in the sperm head with the DNA in the nucleus of the ovum. The sperm and the egg each contribute exactly half of the offspring's hereditary material. The egg undergoes complex maturational changes known as meiosis before it is ready for fertilization. These changes begin during fetal life and are completed only after ovulation and sperm penetration. Table XVI summarizes the comparative aspects of egg and sperm production in relation to fertilization.

The human egg contains 46 chromosomes. This "diploid number" consists of 22 pairs of autosomes and one pair of sex chromosomes. Before fertilization, the number of chromosomes in both eggs and sperm is reduced by one-half to establish the "haploid" number.

Fertilization can only occur within a limited period in each cycle of the limited life span of the sperm and egg (see Chapter 8). The fertilizing life span of sperm is about half its motile life span. Considering the short period of fertilizability of the egg, it can be assumed that the chances of conception are very low 3 to 4 days after the thermal rise of the basal body temperature (BBT) in the female.

Oviductal fluids seem to play a role in normal fertilization. However, early stages of fertilization of the human oocytes have been achieved *in vitro* in the absence of the oviduct. Pregnancy was also achieved in the uterus in the absence of the oviduct following salpingectomy and attachment of the ovary directly to

Table XVI

Physiological Phenomena of Sperm and Egg Related to Fertilization

Parameters	*Oogenesis and Characteristics of Ova*	*Spermatogenesis and Characteristics of Sperm*
Mitosis in the gonad	Mitosis in ovary ceases during fetal life; no new eggs are formed after birth	Mitosis continues throughout reproductive life of the male
Meiosis in the gonad	Begins during fetal life	Begins at puberty and continues throughout reproductive life
First maturational division in gamete	First maturational division is completed in preovulatory follicle	First meiotic division results in two cells of equal size
	A vesicle (first polar body) is pinched off within zona pellucida	Second division follows quickly; 4 spermatids form and mature into 4 sperm
Second maturational division	Second maturational division is completed only when egg is penetrated by sperm	Not comparable
Number of gametes produced during reproductive life	350,000-400,000 oogonia	200 million sperm are produced in each ejaculate from puberty, with reduced numbers during senility
Sex chromosome in gamete	X	X or Y

Amount of cytoplasm in gamete	As oocyte matures, the amount of its cytoplasm increases	As spermatid develops into sperm, the amount of cytoplasm decreases
		Acrosome and tail develop in late spermatid
Dimensions of mature gamete	150 μ	Head length, 4.5 μ; head width, 3 μ; tail length, 55 μ
Motility of gamete	Oocytes, surrounded by follicular cells, are immotile	Sperm motility develops gradually in various parts of epididymis and increases at ejaculation
Plasma membrane at fertilization	Egg acquires plasma membranes from sperm	Sperm loses its plasma membranes to egg
Survival in female reproductive tract	12-24 hr after ovulation	Fertilizability is maintained 24 hr after intercourse and ejaculation

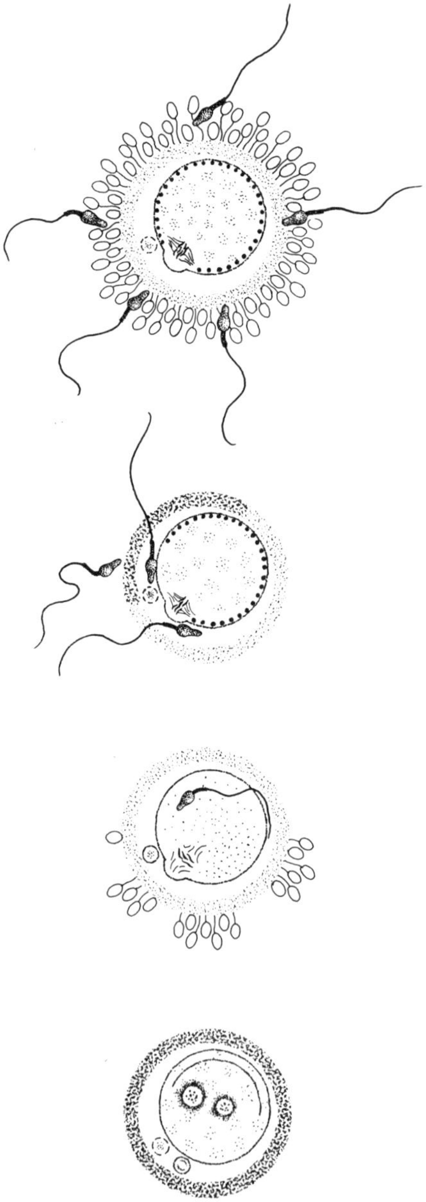

A capacitated sperm reach zona sellucida

B sperm penetrates zona within 10 min.

C sperm penetrates vitellus 2nd polar body to be released

D sperm head forms male pronucleus, vitellus shrinks, a slit is left in zona 2nd polar body extruded

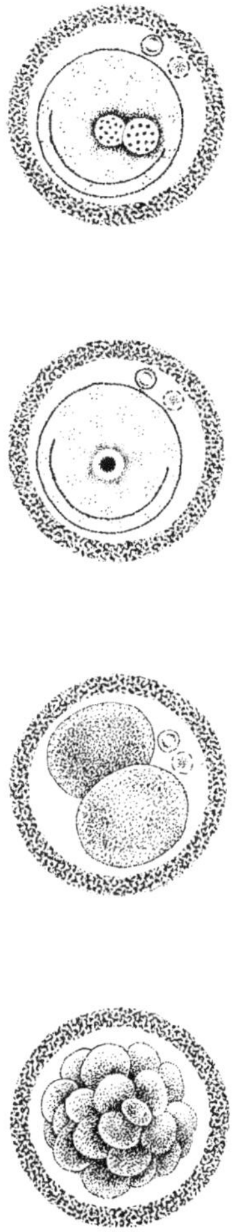

Figure 39. Sequence of events in fertilization, cleavage and early embryonic development in the oviduct. The sperm (head and tail) penetrates the zona pellucida on a slightly curved channel. The sperm head swells to form the male pronucleus whereas the tail of the sperm is detached. The female pronucleus forms around the haploid number of chromosomes. The larger pronucleus is the male pronucleus. During cleavage, the nuclei of the resultant blastomeres have the diploid number of chromosomes. (Adapted from the literature.)

the uterine opening (Estes operation). The literature shows that few pregnancies have occurred following Estes' operation, due to poor fertilization rate in the uterus. This suggests that the fluid may not be vital.

Several unknown biophysical, biochemical and/or physiological factors facilitate the encounter between sperm and egg. Physiological mechanisms ensure entry of only one sperm in the egg during normal fertilization.

Prerequisites and Criteria of Fertilization

There are several prerequisites for fertilization:

1. nuclear and cytoplasmic maturation of gametes in the gonads,
2. development of the fertilizability of gametes in the male and female reproductive tracts;
3. an optimal number of fertilizable sperm with vigorous motility transported to the site of fertilization near the ampullary-isthmic junction of the oviduct; and
4. a fertilizable ovum with a first polar body, transported through the ampulla.

Criteria of fertilization include:

1. penetration of sperm through egg membranes;
2. activation of the egg;
3. formation and syngamy of female and male pronuclei; and
4. first cleavage of fertilized egg.

Capacitation of Sperm

Capacitation refers to various morphological, physiological, and biochemical changes which the sperm undergo during their transport through the female reproductive tract before the sperm are capable of penetrating through the cumulus cophorus, corona radiata and zona pellucida of the egg. Capacitation, a prerequisite for fertilization in several mammalian species, has been assumed to occur in man. Human eggs have been fertilized successfully *in vitro* with ejaculated sperm washed and suspended in artificial culture media with added calf sera.

Capacitation of sperm is associated with several reactions, *e.g.*, changes in sperm membrane near the acrosome, and increased permeability of sperm membrane or increased number of aqueous channels crossing the membrane. Capacitation may involve one or more specific agents in the sperm produced by its aging or by a precise sequence of maternal conditions. Increased metabolic activity of sperm in the presence of oviductal fluids may be a component of capacitation. Follicular and/or endometrial fluid may also be involved in the process of sperm capacitation.

Sequence of Events of Fertilization

Several physiological events are involved in fertilization (Figure 39):

1. Sperm stored in the epididymis undergo maturation. At intercourse and ejaculation, sperm are transported from the site of ejaculation to the site of fertilization.

2. At the time of fertilization, the ovum is surrounded with corona radiata and cumulus cells. These corona cells seem to nourish the egg through channels which reach the vitellus (cytoplasm of egg) through the zona pellucida.

3. The acrosome of the sperm head contains several enzyme complexes, including hyaluronidase and the trypsin-like enzymes which enable the sperm to penetrate through the zona pellucida into the perivitelline space (Figure 40). Incorporation of the sperm into the egg involves the following reactions:

a. membrane fusion; and
b. nuclear reconstruction
   i. pronucleus formation
   ii. exclusion of polar body.

4. Once the sperm penetrates the zona pellucida and the vitellus of the egg, two phenomena take place to prevent polyspermy (penetration of additional sperm in egg): the "zona reaction" and the "block to polyspermy."

5 The entire sperm (head, midpiece and tail) enters the vitellus of the egg. Upon entering the egg, the sperm head swells to form the "male pronucleus."

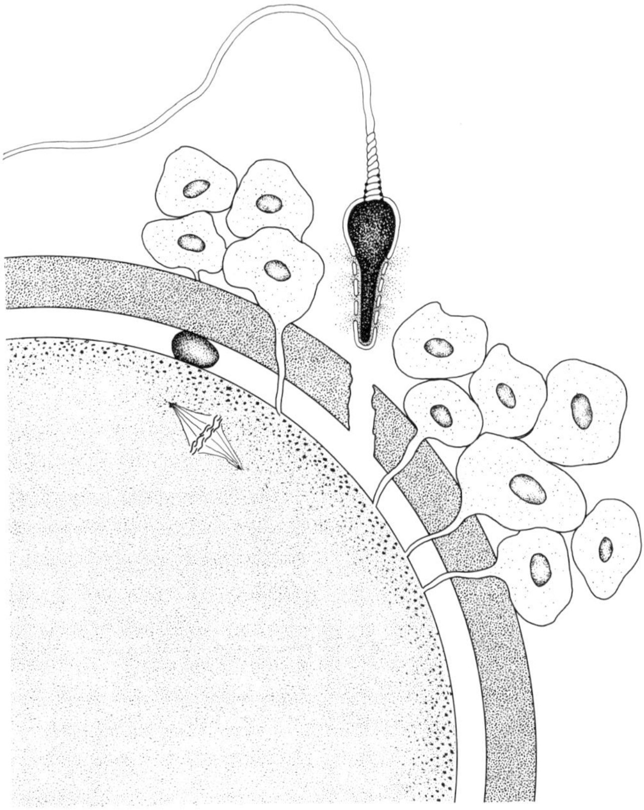

Figure 40. Diagrammatic illustration showing some of the physiological events that take place when capacitated sperm penetrate the cells of the cumulus oophorus in an attempt to reach the zona pellucida, the perivitelline space and the vitellus of the egg. The outer acrosomal membrane of the sperm fuses with the cell membrane at multiple sites creating openings through which the enzyme-rich contents of the acrosome escape. This process leads ultimately to complete loss of the cell membrane over the anterior half of the sperm head. (Adapted from the literature.)

6. The second polar body is extruded from the egg which has a haploid number (23) of chromosomes in the "female pronucleus." The sperm head also contains a haploid number (23) of chromosomes. The "diploid" number (46) of chromosomes is completed when the male and female pronuclei fuse to form the zygote; this process is called "syngamy."

7. After fertilization, the vitellus shrinks and a perivitelline space is formed between the zona pellucida and the vitelline membrane in which two polar bodies are situated.

8. The first cleavage occurs 12 to 18 hours after sperm penetration. During cleavage, the zygote divides into 2, 4, 8 and 16 cells, forming a "morula," which eventually develops into an early "blastocyst" with a blastocoelic cavity (Figure 41).

9. The morula and the blastocyst utilize energy and metabolites from their own vitellus and from the luminal fluids of oviduct.

Aging of Gametes and Chromosomal Anomalies

Unlike other mammals, intercourse and ovulation in humans occur independently. If the time interval between successive intercourses exceeds the fertilizing life of the sperm, it is possible that aged sperm will survive in the oviduct until the time of ovulation. This possibility increases as the couple ages since frequency of intercourse decreases with aging. Also, with advanced age there is a general aging of the genetic material in both the somatic and germinal cells.

The frequency of reproductive failures, chromosomal anomalies, and mongolism increases with advancing maternal age (see Chapter 11). This increase may be due to general aging of the genetic material, overripeness of the ova, or aging of the female reproductive tract.

Normally, the ovum (X) is fertilized by (X) sperm or (Y) sperm. In abnormal cases of nondysjunction of oogenesis or spermatogenesis, fertilization may occur between an abnormal sperm and/or abnormal egg resulting in various chromosomal anomalies (Figures 42 and 43): XO Turner's syndrome or XXX Triple X.

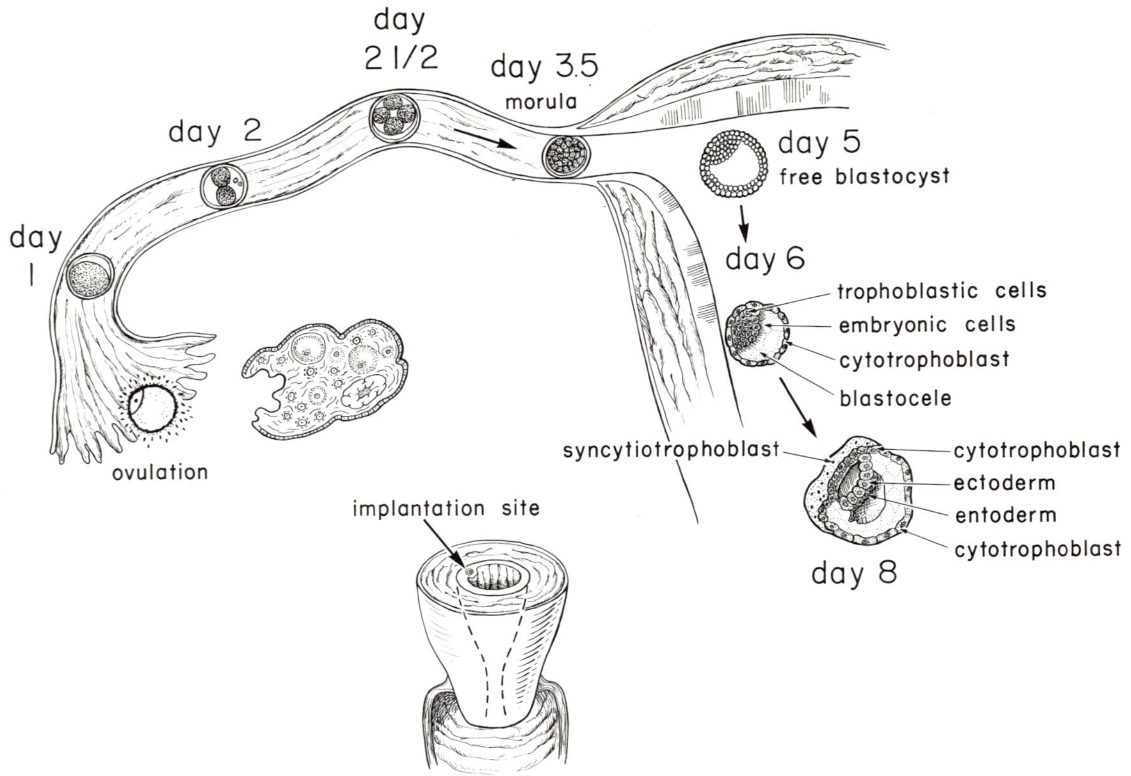

Figure 41. Transport of the fertilized egg in the oviduct and implantation of the blastocyst in the uterus. The arrival of the blastocyst in the uterus coincides with synchronous progestational changes in the endometrium, which occur under the influence of progesterone secreted by the corpus luteum. (Adapted from the literature.)

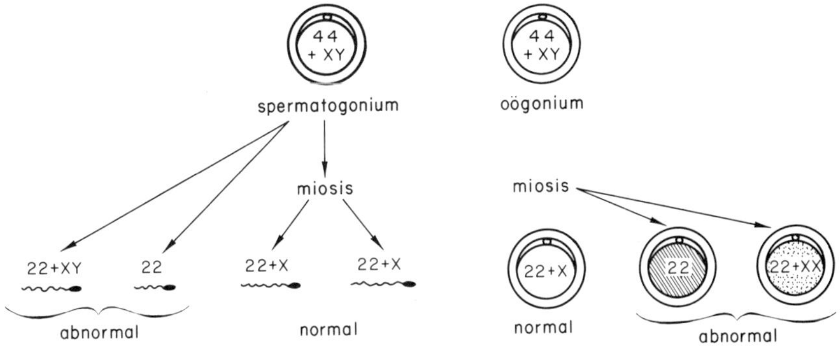

Figure 42. Normal and abnormal spermatogenesis and oogenesis leading to production of normal and abnormal gametes.

Low Temperature Storage and *in vitro* Fertilization of Ova

Ova freshly recovered from the oviduct of laboratory mammals have been stored at -196°C, then thawed and used for fertilization *in vitro* or *in vivo*. Such fertilized ova, on transfer to appropriate recipient, develop to full term. It is possible that these procedures could be applied in clinical medicine. Fertilization of the human ova may take place when the ova are cultured *in vitro* with sperm which has been incubated at least four hours in the uterus of the midcycle rhesus monkey.

CLEAVAGE AND EMBRYONIC DEVELOPMENT IN OVIDUCT

The fertilized egg remains in the oviduct for three days before it is transported to the uterus. Oviductal secretions are necessary for the maintenance and survival of the fertilized egg, since its cytoplasm does not contain sufficient nutrients for cleavage (division of fertilized egg). The nutritional requirements of the fertilized egg seem to vary at various stages of cleavage. Also, the biochemical composition of the oviductal fluid is very different from that of the endometrial fluid. Thus, the premature

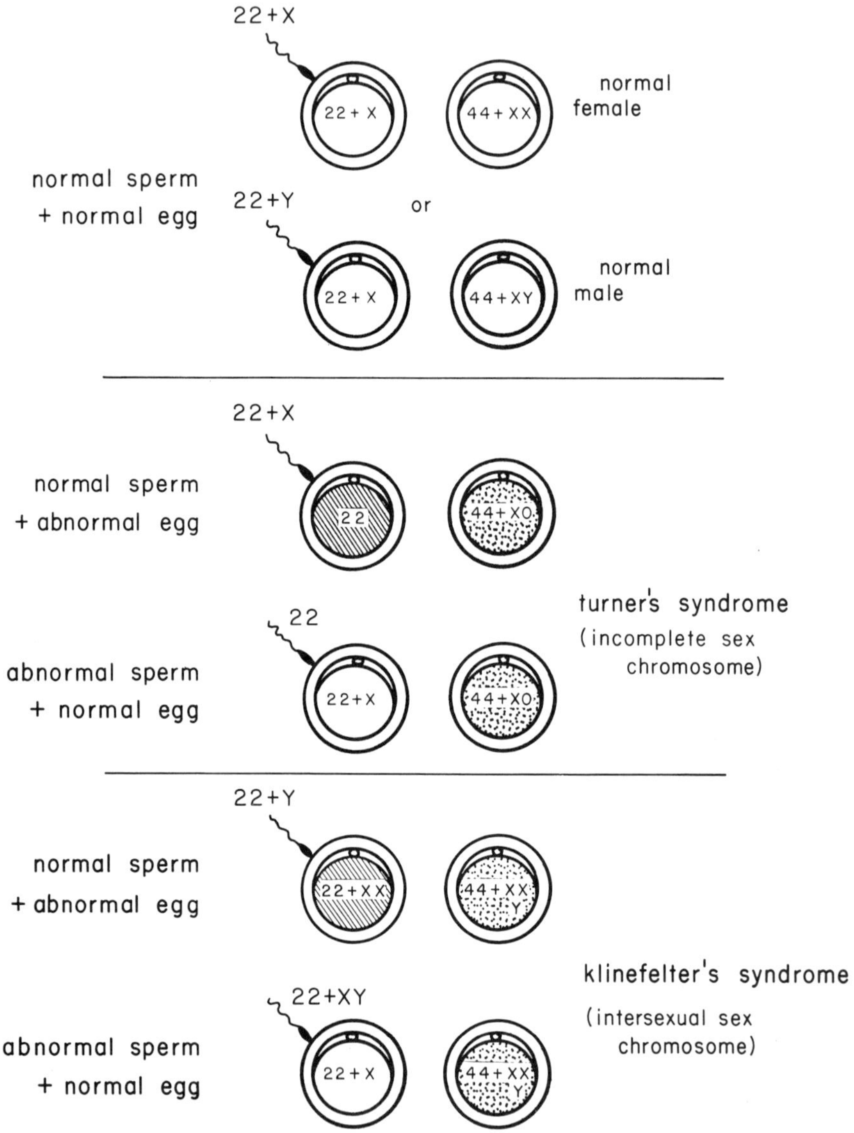

Figure 43. Normal and abnormal fertilization leading to reproductive failure of cytogenetic origin and chromosomal anomalies.

Table XVII

Sequence of Physiological Events Leading to Implantation

Days after Fertilization	*Physiological Events*
3-5	Morula, or early blastocyst, is transported into uterine lumen
4	Morula develops into early blastocyst
5	Blastocyst is composed of 108 cells; zona pellucida remains intact allowing selective transfer of metabolite to embryo during cleavage
6-7	Zona pellucida thins out to allow the trophoblast to slough off or "hatch" (as grape skin is shed off a grape)
8	Trophoblastic cells penetrate the endometrial cells and migrate through its stroma
9	Stickiness of the embyronic trophoblast helps to orient the embryo so that invasion of the endometrium occurs in a precise pattern
10	Cytolysis of and erosion of endometrial tissues occur in response to secreted proteolytic enzymes
11	Synctyiotrophoblast (a synctytium is a multinucleated protoplasmic mass without cell boundaries) forms a thick trophoblastic shell which invades the endometrium

entry of the morula into the uterus will cause its degeneration. After three days, the blastocyst must be transported to the uterus for further development and implantation (Figure 41).

Blastocyst development, implantation and embryonic growth can also take place in the oviduct resulting in tubal pregnancy. Tubal pregnancy, occurs as a result of infection of oviductal mucosa, failure of egg transport, or abnormal muscular activity of the oviduct.

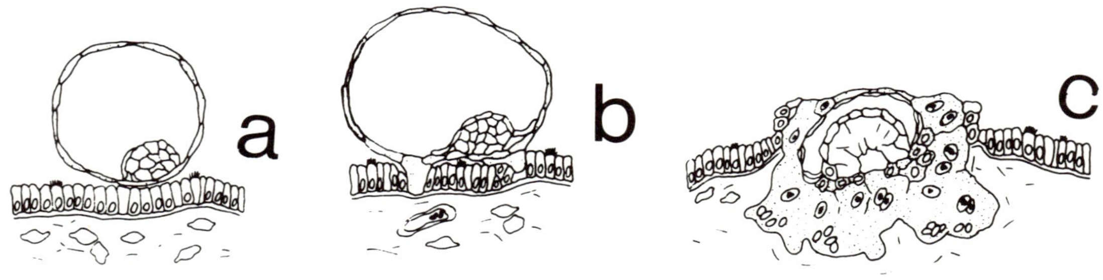

Figure 44. Schematic drawings of stages during implantation of human blastocyst:
(a) adhesion of blastocyst of endometrium;
(b) penetration of trophoblast through endometrium;
(c) prelacunar stage of trophoblast proliferation; and
(d) early lacunar stage indicating vascular invasion and proliferation of cytotrophoblast (Enders *Res. Reprod.* 8(5) 1976).

IMPLANTATION

Implantation of the blastocyst occurs normally in the corpus segment of the uterus, somewhat more often on the posterior wall than on the anterior, and much more often near the midline than laterally. As the fluid begins to accumulate in the morula, the blastocyst is formed with a blastocele (blastocyst cavity) (Table XVII). The cells of the blastocyst differentiate into two cell types: embryonic disc (inner cell mass) and trophoblast. The embryonic disc at one end of the blastocele forms the fetus. The trophoblast invades the endometrium to form the placenta and other extra-embryonic structures which support the fetus during pregnancy. During the course of its differentiation, the trophoblast becomes variously specialized (Figures 41 and 44).

Before implantation, the blastocyst is released from the zona pellucida. There are also dramatic changes in the collagen, cells and blood capillaries of the uterus. The collagen bundles split, disaggregate and degenerate. The endometrial cells undergo hypertrophy and hyperplasia. The blood capillaries in the uterus proliferate and expand.

The metabolism of the blastocyst involves energy production and protein synthesis. This involves developmental changes in enzyme systems. With the initiation of implantation, the uterine contents signal the pituitary gland to secrete a luteotrophic substance which maintains the corpus luteum necessary for early stages of pregnancy. The blastocyst may degenerate before implantation. In such cases, the terminated pregnancy does not affect the length of the menstrual cycle.

Experiments were conducted by transferring embryos from young donor animals to aged recipients, from young to young, and from aged to young. The lowest percentage of embryonic survival was achieved when young embryos were transferred to aged mothers, indicating that poor implantation in aged females is due to defective maternal environment.

SECTION III

REPRODUCTIVE LIFE CYCLE

CHAPTER 10

SEXUALITY

Sexuality, which plays a major role in human existence, involves complex behavioral mechanisms to ensure sexual pleasure from various zones of the body. Sexuality involves sexual identity and sexual behavior (Table XVIII). Sexual identity, the individual sense of maleness or femaleness, is related to the inherited biological sexual characteristics, *e.g.*, hormones, gonads and accessory sex organs. Gender identity refers to the psychological aspects of behavior of masculinity or feminity. Gender is a social concept, whereas sex is a biological one. Although both concepts are congruent, they may develop in conflicting or opposite fashions. Sexual behavior is influenced by the degree of promiscuity, moral standards, and marital relationships and fidelity (Figure 45).

COPULATORY MECHANISMS

Physioanatomical Aspects

Penis

The penis, for most men, is the main source of erotic sensation during intercourse and masturbation. The penis is composed primarily of erectile tissue which becomes engorged with blood during erection. The size of the erected penis is neither related to body conformation, nor affected by sexual experience or masturbation. There have been several myths and fallacies that have related sexuality to the size of the penis. There is no

Table XVIII

Physiological and Behavioral Components of Normal and Abnormal Patterns of Sexuality[a]

Components	*Normal Patterns*	*Abnormal Patterns*
Biological Sex	XX or XY chromosomes program undifferentiated gonad to be testes or ovary which secrete hormones to determine the fetus to be male or female Primary and secondary sexual characteristics are manifested	Anatomical and physiological anomalies (*e.g.*, androgen insensitivity) may cause intersexuality, which may or may not be corrected
Sexual Gender	Positive and reassured sense of femininity or masculinity manifested at 3-4 years of age	Conflicts about femininity or masculinity may cause sex deviations, *e.g.*, homosexuality, transvestism, exhibitionism, or voyeurism
Gender Identity	Persistent inner feeling of femaleness or maleness, as reinforced by reaction of parents, peers and friends	Ambiguous sexual differentiation, morphology and transsexuality, cause rejection of one's body appearance and social status
Sexual Behavior	Desire for sexual pleasure with masculine and feminine connotations leading to orgasm	Abnormal sexual relationships due to faulty communication may lead to complex sexual patterns, conflicts and disturbed perception

[a] [Adapted from H. I. Lief. *The Sexual Experience,* B. J. Sadock, H. I. Kaplan and A. M. Freeman, Eds. (Baltimore: Williams & Wilkins, 1976).]

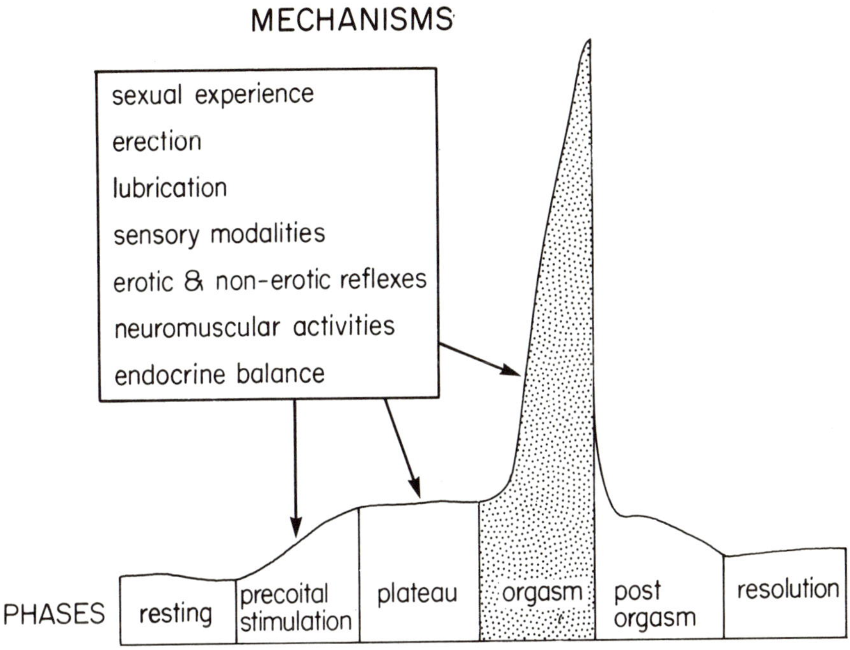

Figure 45. Diagrammatic illustration showing the physiological mechanisms that influence the various stages of sexual behavior.

evidence to indicate that a large penis is more satisfying to a woman, and that a circumcised penis is more sensitive and therefore more liable to premature ejaculation than an uncircumcised one.

Vagina and Vulva

The vagina is a highly distendable yet adaptable organ. The penis acts as an appropriate dilator, bringing the vagina to conformational shape. The inner (upper) vagina is sexually insensitive, whereas the outer part of the vagina, the vulva, and the clitoris are sexually excitable. Two centers of sexual sensory perception are located along the vaginal walls at approximately the 4:00 and 8:00 o'clock areas, approximately two-finger joints inside the vagina.

The hymen may exist intact as a covering for the vaginal orifice until intercourse, or it may be broken or pierced by accident or experimentation. Most hymens before intercourse will allow passage of a vaginal tampon or inspecting finger whereas others may stretch quite easily.

The size of the clitoris varies considerably and is unrelated to the degree of sexual responsiveness of a particular woman. During sexual excitement, the clitoris does not undergo erection, but becomes engorged with blood and increases in sensitivity. The glans clitoris contains the highest number of pressure receptors in the body. The clitoris is not necessarily the prime focus of sensitivity for the female.

Mammary Gland

The size, consistency and conformation of a woman's breast plays a role in sexual attraction. Micromastia (small breasts), macromastia (large breasts), bilateral asymmetry and inverted nipples are developmental in origin. Chronic pulmonary or cardiac diseases may be affected by macromastia. In men, gynecomastia (development of breasts) often accompanies liver diseases and alcoholism.

Hormones and Sexual Responses

The maintenance of sexual response in men depends to a large extent upon androgen secretion, whereas sexual behavior of women is relatively independent of ovarian hormones. Castration of the adult male will, after a latent period of several months, result in impotence and a decline of aggressive behavior. Ovariectomy does not usually affect sexual behavior in women who have acquired the learned habit of pleasurable sexual response. Also, in otherwise normal women complaining of frigidity, there are no abnormalities of ovarian hormone production.

Copulatory Patterns

In general, men show less variability from one to another and from one time to another in their sexual patterns and responses than women, who are labile and variable. Several coital techniques have been adopted:

1. Genital-genital position
   a. Face-to-face position
      i. man on top
      ii. woman on top (female superior)
   b. Side-by-side position (lateral)
   c. Sitting and standing
   d. Rear penetration
2. Genital-oral position
   a. Cunnilingus
   b. Fellatio
3. Hand-genital position
   a. Female masturbation
   b. Male masturbation
4. Penis-anus position

The most common position is the male superior/female supine position. The side-by-side position and the rear penetration position are common during late pregnancy.

Stretching and pressure of the mons and labia induce erotic sensation. The clitoral prepuce is contiguous with the labia minora. During intercourse, there is no direct stimulation of the clitoris by the penis. However, penile thrusting exerts adequate traction on the labia to induce orgasm. Although there is no direct contact of the clitoris by the penis, the labia minora exerts pressure on the clitoris (Figures 46 and 47). For better mobility of her pelvis, the woman keeps her legs bent and at the sides of her partner or entwines her legs around the man's back rather than his shoulders. Most women prefer stimulation of the clitoris, particularly the shaft of the clitoris, since excessive stimulation of the glans of the clitoris causes a particular type of hypersensitivity.

Duration, Frequency and Intensity of Intercourse

The duration of intercourse varies from 5 to 20 minutes according to race, age, environmental conditions, physical fitness, disease, drug therapy and the type of relationship between the partners. For example, intercourse is of short duration for bachelors with heightened precoital tension. Gradual changes are noted in sexual responses in both sexes during aging.

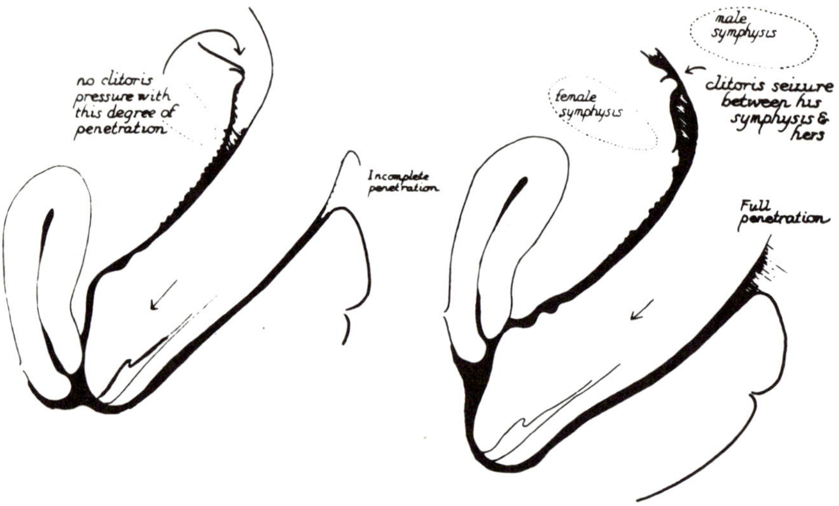

Figure 46. At full penetration, the clitoris may (right) or may not be stimulated (left) by male-female symphysis pressure rather than by the penis itself [From Dickinson, R.L. *Atlas of Human Anatomy* (Baltimore: Williams & Wilkins, 1949).]

The mean frequency of intercourse in American married couples is 4.8/week. This average declines to 1.8/week in the 50s and 1.3 in the 60s (Figure 48). The mean frequency of intercourse in Africans is higher than those reported for Americans.

In laboratory animals and domestic farm mammals, coitus is normally restricted to the period of estrus, which is synchronized with ovulation. In human and nonhuman primates, the female may be sexually receptive throughout the menstrual cycle. Thus in these species, sexual receptivity and ovulation are not synchronized. In monkeys, maximal sexual activity occurs during and immediately before the expected time of ovulation. Sexual receptivity of women is not confined to the period of ovulation, and intercourse occurs frequently at times other than the fertile period.

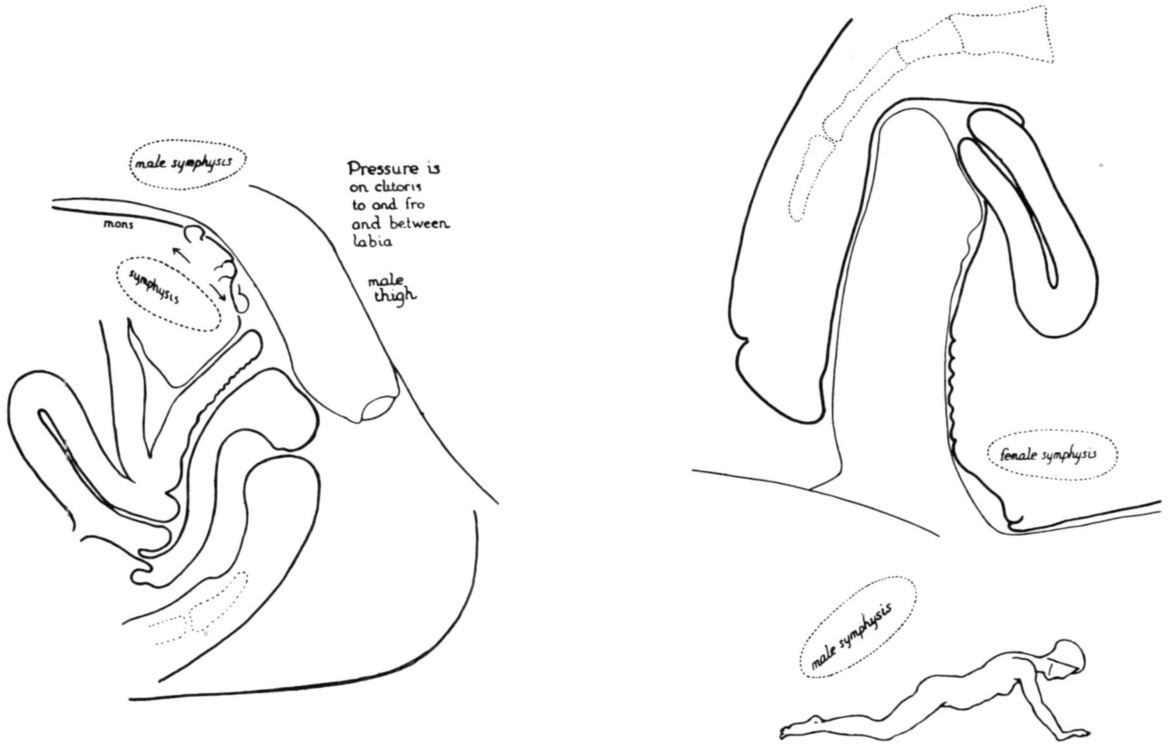

Figure 47. *Left:* The flaccid penis is in contact with the vulva. Note the excursion of the clitoris in the vertical plane, which is produced by pelvic thrusting and which may lead to female orgasm.
Right: In this position, the woman lying above can regulate the rate and depth of penile penetration. Maximal clitoral stimulation is provided in the woman on top position [From Dickinson, R. L. *Atlas of Human Anatomy* (Baltimore: Williams & Wilkins, 1949).]

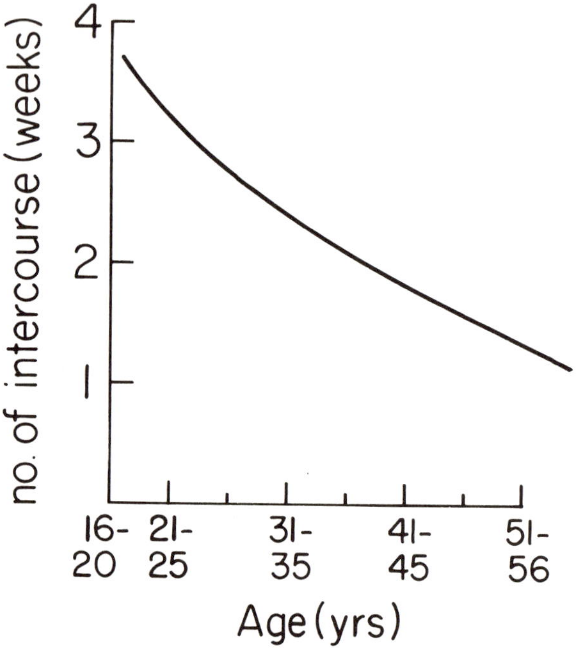

Figure 48. Effect of age on frequency of intercourse in men. Note the steady decline in sexual desire with aging.

SEXUAL BEHAVIOR IN MEN

Sexual response in men includes precoital excitement, penile erection, plateau phase, ejaculation associated with orgasm and a refractory phase (Table XIX). These responses are regulated by various parts of the central nervous system.

Precoital Excitation

Sexual arousal in men is initiated by various intracerebral mechanisms, such as thoughts, sexual fantasy and sensory stimuli, *e.g.*, visual, auditory, olfactory or tactile perceptions. Precoital excitement causes penile erection and muscular tension. The testes and scrotum are pulled closer to the body causing thickening of scrotal skin and shortening of the spermatic cord.

Penile Erection

The transformation of the penis from a flaccid to an erect condition is a vascular phenomenon. The penis consists of three parallel tubes: two corpora cavernosa lying side by side, and the corpus sponigiosum which surrounds the urethra. These structures are made up of sponge-like erectile tissue with vascular spaces interspersed between the arteries and veins.

Activation by the parasympathetic nervous system stimulates the blood vessels of the area to dilate. The vascular spaces become distended with blood at high pressure causing noticeable increases in penile size. The contraction of ischiocavernosus and bulbocavernosus muscles helps pump blood to the penis and to maintain erection by mechanical obstruction of the venous drainage from the penis. The rate of arterial inflow becomes temporarily greater than that of venous outflow.

The length of the penis ranges from 7 to 11 cm (average 10.5) in the flaccid state and from 14 to 18 cm (average 15.5) in the erect state. The dimensions in the flaccid state are unrelated to the erect dimensions. Thus, the smaller penis erects to a proportionally greater size than does the larger penis.

There are two types of penile erection: psychogenic (cerebral) and reflexogenic. Psychogenic erection is provoked by stimulation of peripheral nerves and the erotic center of the brain in response to auditory, visual, olfactory, gustatory, tactile, or imaginative stimuli. Reflexogenic erection results from exteroceptive stimulation of the external genitalia (*e.g.*, lightly stroking the glans penis), or interoceptive stimulation in the rectum and/or bladder. Most erections that occur during sleep are associated with dreams or are reflex responses to visceral or tactile impulses.

Dysfunctions in penile erection may be due to various anomalies affecting the afferent signals in thoracolumbar or sacral erection centers, or the vascular supply or the erectile tissues of the penis. Penile erection is weakened, and ejaculation is delayed as a result of several factors, *e.g.*, drug therapy (antidepressants, antihypertensives, tranquilizers), alcohol, marijuana, depression, fear of failure, anemia, diabetes, fatigue, obesity, malnutrition and various metabolic disturbances.

Table XIX

Patterns of Normal Sexual Response in Men and Women

Phases	*Men*	*Women*
I. Precoital Excitement (several min-hr)	Penis: erection within 10-30 sec caused by vasocongestion of erectile bodies of corpus cavernosa; loss of erection may occur with introduction of asexual stimulus, loud noise Testes: partial elevation toward perineum Scrotum: tightening and lifting	Breasts: nipple erection; venous congestion; areolar enlargement Clitoris: diameter of glans increases two-fold Labia majora: elevation and flattening against perineum in nullipara; rapid congestion, edema and increase in size in multipara Labia minora: color change to bright pink or red; increase in size Vagina: vaginal transudate within 10-30 sec; drops of clear fluid coalesce for lubrication of vaginal barrel (aids in buffering acidity of vagina to neutral pH required by sperm); *Upper two-thirds:* balloons and dilates convulsively as uterus moves up pulling anterior vaginal wall with it; fornices lengthen, rugae flatten; *Lower third:* dilation of vaginal lumen, congestion of vaginal walls Uterus: ascends, moves into false pelvis Cervix: passively elevated with uterus

II. Plateau (30 sec-3 min)	Skin: sexual flush Penis: increase in size of glans and diameter of penile shaft Testes: increase in size due to vasocongestion; flattening of testes against perineum signals impending ejaculation Cowper's glands: 2-3 drops of mucoid fluid containing viable sperm Blood pressure and respiration: increase Semispastic contractions of facial, abdominal and intercostal muscles	Skin: sexual flush inconstant except in fair-skinned; pink mottling on abdomen, spreads to breasts, neck, face, often to arms, thighs and buttocks Breasts: increase in size, especially in breasts that have not been nursed; enlarged areolae Clitoris: shaft withdraws deep into swollen prepuce just before orgasm Labia majora: enlarged and edematous (hang like folds of a heavy curtain) Labia minora: enlarged labia gape widely to form a vestibular funnel into vaginal orifice; orgasm will then always follow within 3 min, if stimulation is continued Bartholin's Glands: a few drops of mucoid secretion form; aid in lubricating vestibule (insufficient to lubricate vagina) Vagina: copious transudate continues by prolonged preorgasmic stimulation *Upper two-thirds:* further ballooning creates diameter of 2-3 in.; wall relaxes in a slow tensionless manner

Table XIX, continued

Phases	*Men*	*Women*
II. Plateau (continued)		*Lower third:* maximal distension reached rapidly; lumen contracts to half its diameter; contraction around penis allows thrusting traction on clitoral shaft via labia and prepuce Uterus: strong sustained contractions (lasting 2 min) similar to that during late labor Cervix: slight swelling; patchy purple
III. Orgasmic (3-15 sec)	Skin: well-developed flush Penis: ejaculation; 3-4 contractions at 0.8 sec of vas, seminal vesicles, prostate and urethra, followed by minor contractions with increasing intervals	Vagina: transudate provides maximal lubrication *Upper two-thirds:* fully ballooned out and motionless *Lower third:* 3-15 contractions of lower third and proximal labia minora at ¾-sec intervals Uterus: contractions throughout orgasm; strongest with pregnancy and masturbation Rectum: rhythmical contractions particularly with masturbation rather than with intercourse External urethral sphincter: occasional contraction, without urine loss Irregular spasms

IV. Resolution (10-15 min; if no orgasm, 0.5-1 day)	Skin: flush disappears in reverse order of appearance Penis: erection–partial involution in 5-10 sec with variable refractory period; full detumescence in 5-30 min Scrotum and Testes: decrease to baseline size due to loss of vasocongestion; testicular and scrotal descent 5-30 min after orgasm; involution may take several hours if there is no orgasmic release; return to baseline state in 5-10 min	Skin: fine perspiration, mostly on flush areas Clitoris: shaft returns to normal position in 5-10 sec; full detumescence in 5-30 min (if no orgasm, clitoris remains engorged for several hours) Labia majora: increase to normal size in 1-2 min in nullipara; decrease to normal size in 10-15 min in multipara Labia minora: clitoral detumescence (as rapidly as loss of erection in men) Vagina: some transudate collects on floor of upper two-thirds formed by its posterior wall (in supine position); ejaculate deposited in this area forming seminal pool *Upper two-thirds:* cervical descent–descends to seminal pool in 3-4 min *Lower third:* congestion disappears in seconds (if no orgasm, congestion persists for 20-30 min) Uterus: descends and returns to normal Cervix: color and size return to normal in 4 min

[Data from V. A. Sadock, after Masters & Johnson data; from *The Nature and Evolution of Female Sexuality,* M. J. Sherfey, Ed. (New York: Random House, Inc., 1966,1972).]

Ejaculation

During sexual development, the capacity for ejaculation appears much later than erection; adolescent boys may experience nocturnal emission without erection. Ejaculation occurs in two stages as follows:

1. Ejaculatory inevitability: this stage is of brief duration, 2-4 seconds, and is difficult to control. Continued stimulation of the glans penis causes a summation of efferent impulses until a specific threshold is reached. There is considerable variation in the threshold at which ejaculation occurs among men and even periodically in the same individual.
2. Emission of semen: sudden contraction of the seminal vesicles is associated with reflex contraction of bulbocavernosus and ischiocavernosus muscles, causing the discharge of spurts of semen, which are propelled through the full length of the penis.

The reflexes of emission and ejaculation, elicited by stimulation of the genital organs, can be intentionally initiated, but once started cannot be influenced voluntarily. The process of ejaculation is regulated by psychological factors and the sympathetic nervous system. The psychological aspect of ejaculation may involve concentration on the present partner or act, or fantasy about other times and persons.

Orgasm

Orgasm, not identical to sensation of ejaculation, is generally a local and pleasurable sensation associated with gratification and relief of tension at climax. Orgasm is triggered by the sensation of ejaculatory inevitability, followed by the forceful emission of semen. Orgasm is associated with four to five spasms of the prostate, seminal vesicle, vas deferens and urethra, and involuntary contractions of the internal and external anal sphincter. Orgasm is also associated with remarkable increases in respiration rate, oxygen consumption, pulse rate, blood pressure and muscular tension. These changes, comparable to moderate rather than strenuous exercise, are quite brief, and are followed by a drop to normal levels within a few minutes after orgasm, and rapid release from vasocongestion and myotonia (muscle tension). Maximal orgasmic potential occurs at 18-20 years of age when some eight orgasms may be achieved during a 24-hour period. After the age of 30 years, one orgasm per day is not uncommon.

In certain neurological disorders, ejaculation may be dissociated from orgasm. In such cases, the resolution of the anatomical and physiological changes associated with sexual stimulation become drawn out and less orderly. The recurring incidence of failure to obtain orgasmic release following sexual excitation may lead to psychosexual dysfunction.

Refractory Phase

After orgasm, men become refractory to sexual stimuli when the penis returns to its flaccid state. Erection subsides promptly following ejaculation since the sympathetic impulses which facilitate emission of semen also cause constriction of blood vessels supplying the erectile tissues of the penis. In the nonerect penis, the vascular spaces contain very little blood and are collapsed. The refractory period, which may last from several minutes to many hours, may be associated with sleep.

SEXUAL BEHAVIOR IN WOMEN

The peak of sex drive in women is commonly reached after the age of 25, but a high degree of libido may continue into old age, subject to the availability of a potent partner. During pregnancy, there is a steady and consistent decrease in sexual activity and a decline in the frequency, strength or intensity of orgasm. Sexual desire resumes the normal pattern three months after delivery.

Menopausal and postmenopausal women maintain the multiorgasmic capacity of their younger years. However, in a few postmenopausal women, painful intercourse may result from the thinning of vaginal walls and decreased vaginal lubrication. Burning and frequent urination after intercourse may occur since the atrophic bladder and urethra are more susceptible to irritation and not as protected by the thin vaginal walls. Secondary stenosis or narrowing of the vagina may occur in postmenopausal life and after long sexual abstinence. Several responses decline in response to disease and surgery of the urogenital system, systemic diseases, drug abuse, excessive fatigue, overeating, alcoholism and emotional disturbances.

Four phases are noted during intercourse: precoital excitement, plateau, orgasm and refractory period (Table XIX).

Precoital Stimulation

Visual, tactile, auditory and olfactory stimuli interact centrally to transform the woman from a resting phase to excitement and through plateau and orgasm. Precoital stimulation is associated with nipple erection, vaginal enlargement and lubrication (sweating-like phenomenon of the vaginal wall). During the resting phase, the vagina is some 8 cm in length. With precoital excitement, the vagina undergoes dilation, distention and elongation.

Since the vaginal environment is hostile to sperm survival, a rapid sperm transport into the cervical mucus is necessary for sperm survival in the female reproductive tract. Prolonged precoital stimulation and happy intimacy of the couple enhance the secretion of vaginal fluid which neutralizes vaginal acidity in favor of conception, in cases where the sperm count is marginal for fertility. On the other hand, the absence of precoital stimulation and lack of female orgasm might prolong infertility.

ORGASM

Orgasm is a psychophysiological experience associated with perception of a peak of physical reaction to sexual stimuli, and a brief episode of release from vasocongestion and muscle tension built up during sexual excitement. Orgasm lasts 3 to 15 seconds and is associated with:

1. slight clouding of consciousness;
2. 3 to 12 involuntary contractions of the vaginal orgasmic platform;

3. contractions of the uterus, flowing from the fundus downward to the cervix at intervals of 0.8 seconds;
4. involuntary contractions of the internal and external anal sphincter;
5. vocalization;
6. voluntary and involuntary movements of the large muscle groups, including facial grimacing and foot spasms; and
7. changes in blood pressure, heart rate and respiratory patterns.

Following orgasm, women may resolve toward sleep or remain responsive to sexual stimulation.

Sexual gratification during intercourse is not necessarily accomplished by simultaneous orgasm of the couple since each partner can attend to the pleasure needs of the other. During orgasm, a woman desires deep vaginal penetration and continual and sustained direct or indirect clitoral stimulation, whereas a man tends to withdraw his pelvis and halt thrusting movements.

HOMOSEXUALITY

Homosexuality is a sexual variation within the normal range of psychological functioning. Homosexuality is experienced by 10% of adult men and 5% of women, and is more common in adolescent boys than in adolescent girls. Homosexual persons become erotically aroused by persons of the same sex and/or engage in sexual behavior with persons of the same sex. Homosexual males perform cunnilingus, masturbation or use body friction to stimulate sexual gratification.

Homosexual experiences are common when youth of the same sex congregate. Overt homosexuality may be due to fear of the opposite sex, fear of adult responsibility, a need to defy authority, flight from reality, loneliness, boredom, rebelliousness, curiosity or neurotic need to please.

SEXUAL AND PSYCHOSEXUAL DYSFUNCTIONS

Anomalies of psychosexual responses are not uncommon. Erectile impotence, premature ejaculation, retarded ejaculation or priapism (persistent erection) are the common syndromes in men (Table XX), whereas orgasmic dysfunction, vaginismus (painful spasm) and dyspareunia (painful coitus) are the main symptoms in women.

Psychosexual dysfunction may be due to endocrine disorder, administration of large doses of drugs, anatomical anomalies, systemic or neurological diseases, tumors and other pathological processes in the hypothalamo-pituitary system, past unfavorable sexual experiences, marital problems, or traumatic events at crucial stages of psychic development. Courtship-disturbed males (exhibitionists, voyeurs, toucheurs, frotteurs, rapists and obscene callers) have adequate masculine gender identity and appropriate sex object preference but lack the assertive characteristic of the male sex role. This is particularly evident in the early pretactile stage of courtship.

Sexual Impotence

Sexual impotence refers to the inability to achieve penile erection and failure to accomplish satisfactory intercourse and sexual gratification. Impotence may be "primary" if it occurs from the onset of sexual maturity or "secondary" if the erection has previously been effective. A patient may be selectively impotent in certain situations or with a certain partner.

Various disorders may cause sexual impotence:

1. neurological disorders, which destroy the function of the lower thoracolumbar sympathetic fibers;
2. prostatectomy, particularly perineal prostatectomy;
3. hypoandrogenism associated with pituitary destruction or insufficiency, hypothalamo-pituitary imbalance or hypothalamic lesions;
4. various endocrine disorders, *e.g.*, diabetes mellitus, hypopituitarism, eunuchoidism, myxedema, thyrotoxicosis, Addison's disease, acromegaly and adrenal feminization; or
5. administration of large doses of addictive, psychotropic or antidepressant drugs.

Table XX

Some Patterns of Sexual Behavior and Related Anomalies in Men

Syndrome	*Behavioral Patterns*
Premature Ejaculation	Ejaculation before or within 2 min of penile intromission into the vagina; inability to delay ejaculation long enough to allow partner to achieve orgasm in 50% of episodes of intercourse
Retarded Ejaculation	Difficulty in ejaculation even after intense sexual stimulation; more than 30 min between intromission in vagina and ejaculation
Ejaculatory Failure (dribbling)	Strong ejaculatory contractile thrust replaced by a few weak contractions, and semen dribbles from penis; may result from total paralysis of ejaculatory striated muscles
Priapism	Sustained erection caused by sexual stimulation without pleasurable sensation; may be due to severe brain damage; may be painful and lead to thrombotic changes of cavernous bodies; may be associated with certain hematological disorders, *e.g.*, sickle cell anemia or leukemia
Coitus Interruptus*	Withdrawal of penis from vagina before ejaculation
Coitus Reservatus*	Inhibited ejaculation within or outside the vagina; avoiding too close an approach to orgasm allowing penile detumescence to occur slowly over a prolonged period
Coitus Saxonicus*	Propulsion of seminal fluid from posterior urethra into the bladder with retention of sensation of ejaculation; male exerts pressure of his perineum causing retrograde ejaculation into bladder; ejaculate emitted later during urination
	Involuntary ejaculations in bladder due to closure of internal bladder sphincter

The physiological mechanisms by which diabetes mellitus may cause impotence are obscure. Impotence seems to occur irrespective of the duration or mode of treatment (insulin or oral agents) of the disease. However, impotence is more common in aged diabetic patients and in patients suffering from diabetes for long periods. Certain surgical procedures, such as transurethral resection of the prostate, do not interfere with neural and vascular mechanisms of penile erection.

Impotence can be reduced greatly by a combination of various treatments, such as systematic desensitization and application of erotic audiovisual stimuli.

Ejaculatory Disturbances

Disorders of ejaculation include premature ejaculation and retarded or inhibited ejaculation (Table XX). Premature ejaculation occurs within the first 30 to 60 seconds after intromission, implying the inability to postpone ejaculation within normal limits. Ejaculatory difficulty may be due to several factors; *e.g.*, neurological disorders interfering with the sympathetic innervation to the genitalia, such as lumbar sympathectomy and administration of antihypertensive drugs.

Frigidity (Sexual Anesthesia)

Frigidity is associated with inadequate orgasm or lack of response to sexual stimulation. Frigidity may result from dyspareunia, vaginismus or fear of pregnancy. Primary frigidity, usually of psychogenic origin, requires treatment by a competent team of endocrinologist, psychiatrist and sex counselors. Secondary frigidity from manifold causes is easier to manage.

Failure to achieve orgasm does not necessarily indicate lack of sexual response. Preorgasmic women, those who never experience an orgasm, may be taught the orgasmic sensation through appropriate masturbatory experiences. Other women can repeatedly experience orgasm with masturbation but are unable to experience it with a sexual partner due to problems in communication.

Dyspareunia

Dyspareunia, difficult or painful intercourse, may be felt either deep in the pelvis or superficially at the vaginal introitus. Dyspareunia may result from various conditions:

1. anatomical defects, *e.g.*, a rigid, thick or unruptured hymen, presence of a vaginal septum, or congenital narrow introitus;
2. pelvic inflammation or infection of the vulva, urethra, or glands of Bartholin;
3. laceration of the broad ligament or presence of tumors or cysts;
4. infectious, senile or radiation vaginitis;
5. allergic sensitivity reactions to douches, creams, jellies, or the latex in condoms and diaphragms;
6. postmenopausal vaginal atrophy; or
7. painful scars on the roof of the vagina following hysterectomy, vaginal repair, or delivery.

Vaginismus

Vaginismus is an involuntary spasm of the pelvic musculature and lower third of the vagina. Vaginismus may be due to a tight introitus, a tender vagina or fear of coitus. Diagnosis of vaginismus is based on physical demonstrations of spasms in the perineal muscles. Vaginismus can be treated by self-dilation of the introitus and vagina, systemic deconditioning, Kegal exercises, or behavioral therapy.

Psychosexual Tests

Several neurological diagnostic tests and reflexes are used to determine whether sexual impotence is caused by the disease itself or whether it is a psychogenic reaction: (1) tests of voluntary activity of pelvic floor muscles, (2) genital sensory tests, preferentially testing vibration, and (3) eliciting a number of genital reflexes. Electromyography of pelvic floor muscles, urodynamic studies, tests of rectal muscular function, glucose tolerance tests and spinal fluid protein studies are also helpful. Neuroendocrinological examination includes skull X-ray, computerized tomography or pneumencephalography, visual field on, and estimation of blood testosterone, estradiol, FSH and prolactin.

Behavioral Psychotherapy

Behavioral psychotherapy for the treatment of sexual deviation, deals with overt and objectively observable behavior, rather than with thoughts and feelings. This type of therapy aims at symptomatic improvement and the elimination of suffering and maladaptive habits using various conditioning and anxiety-eliminating techniques derived from learning theory. Behavioral therapy, a new approach to psychotherapy based on learning principles, deals chiefly with problems caused by psychological factors. Modern therapy is based on three concepts: (1) the couple is always treated together; (2) the negative sexual behavior chain is interrupted at the beginning by prohibiting sexual intercourse; and (3) the behavioral chain of events is reestablished step by step. The couple may be instructed about alternative coital postures, and a pattern of sexual stimulation recommended to suit their mutual needs.

Nonorgasmic women are educated about their sexual anatomy and physiology in conjunction with their partners. Women are encouraged and instructed in visual and tactile examination of their external genitals through use of mirrors. Self-exploration of the vagina and palpation of the cervix is also taught.

CHAPTER 11

REPRODUCTIVE FAILURE

Reproductive failure occurs in 10 to 15% of married couples as a result of anatomical, physiological, endocrinological, pathological, immunological or psychosexual factors (Figure 49). One-third of the failure is due to andrological infertility, one-third to gynecological failure, and the remainder is unexplained.

REPRODUCTIVE FAILURE IN MEN

Reproductive failure in men may be linked to a dysfunction interfering with production and release of sperm, secretory activity of the accessory organs or to difficulties in intercourse as a result of anatomical, physiological, endocrine, mechanical, immunological, genetic, vascular, neural, psychogenic, iatrogenic, or pathological factors (Table XXI). Inability to form viable sperm may result from various factors including anomalies in the structure and/or function in one of the male reproductive organs. Inability to fertilize an ovum, even though sperm appear normal, may be caused by undetected chromosomal error or abnormal sperm. Male infertility may cause emotional reactions and subsequent psychosomatic disorders. Aging of the male is accompanied by senile involution of the testis, but reproductive capacity is retained until an advanced age.

Testicular Anomalies

Testicular failure may be primarily of testicular origin or may be due to hypothalamic or pituitary insufficiency. Primary

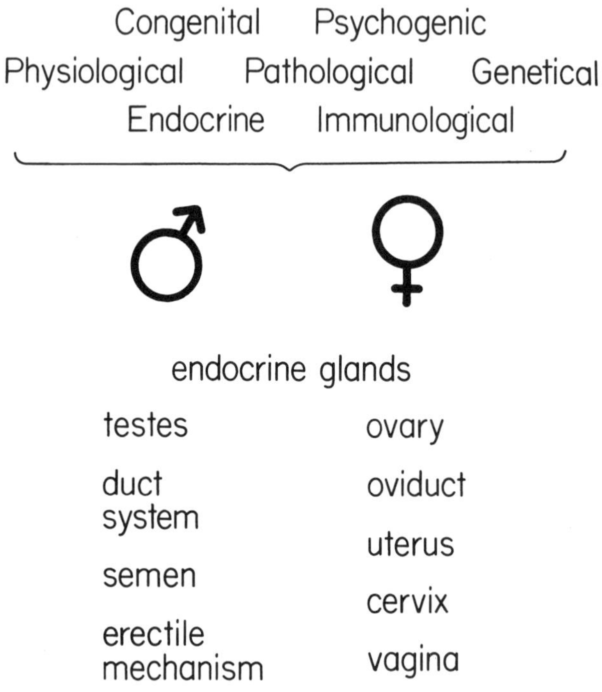

Figure 49. Summary of the congenital, psychogenic, physiological, pathological, genital, endocrinological and immunological factors causing reproductive failure in a couple.

testicular failure may involve: (1) both the germinal and Leydig cell compartments (*e.g.*, testicular agenesis); (2) the germinal epithelium, only as in some cases of Klinefelter's syndrome; or (3) Leydig cell, only as in male climacteric. Primary germinal epithelium resulting from genetic factors can be diagnosed by buccal smear and chromosomal analysis.

Cryptorchidism results from failure of the testis to migrate from the abdominal cavity, probably as a result of either a small, inactive testis, or adhesions between testes and adjacent structures. Unilateral or bilateral cryptorchidism causes defective spermatogenesis because increased testicular temperature causes damage to the seminiferous tubules. The longer the testes remain

Table XXI

Types of Reproductive Failure and Some Methods of Differential Diagnosis in Infertile Men

Types of Disorders	*Reproductive Failure Mechanisms*	*Differential Diagnosis*
Developmental	Testicular-germinal aplasia	Clinical history
	Hypospadias	Physical examination
	Varicocele	Vasography
	Ductal obstruction	Retrograde venography
	Congenital absence of vas deferens	Scrotal thermography
Genetic and Chromosomal	Hermaphroditism Pseudohermaphroditism Klinefelter	Karyotyping (buccal smear, sex chromatin)
Testicular	Cryptordchidism Retarded descent Floating testis Pendulous testis	Testicular biopsy: histology, histochemical and tissue culture of seminiferous tubules
	Testicular hypoplasia (primary or idiopathic)	Scrotal thermography
	Testicular atrophy (trauma, inflammation) Radiation injury (idiopathic)	Ultrasonography

Table XXI, continued

Types of Disorders	*Reproductive Failure Mechanisms*	*Differential Diagnosis*
Endocrine	Testicular insufficiency (primary, secondary)	Endocrine profile: FSH, LH, GnRH
	Pituitary disorder (pre- and postpubertal LH deficiency)	
	Hypothalamic GnRH deficiency	17-hydroxy-steroids, 17-ketosteroids, HCG stimulation
	Thyroid deficiency	
	Diabetes mellitus	
Neuroendocrine and Psychological	Disorders of erection	Psychological tests
	Disorders of ejaculation	Radiology (cervical vertebrae, lumbar vertebrae)
	Impotence	
Semen	Disorders in: sperm concentration; sperm morphology; sperm motility	Semen analysis: counting, size, motility (live/dead %) (% abnormal sperm)
	Liquefaction of semen coagulum	
Infection	Gonorrhea, syphilis	Incubation of semen culture
	Mycoplasma, trichomonas, Candida, orchitis, epididymitis, prostatitis, urogenital tuberculosis	
Immunological Disorders	Sperm agglutination	Agglutination, microagglutination
		Immunofluorescence of sperm

Impotence	Acromegaly, diabetes mellitus, eunuchoidism, myxedema, obesity	Endocrine functions
	Multiple sclerosis, spina bifida, Parkinsonism	Neurological tests
	Priapism, phemosis, perineal prostatectomy	Urological tests
	Drugs: amphetamines, atropine, guanethidine, impiramine, monosamine, oxidase inhibitor, phenothiazines, reserpine	
General	Fatigue, excessive smoking, alcohol, excess of intercourse, fear, aging, allergic reactions, nutrition	Medical history

bdomen, the greater the testicular fibrosis. If the not descend within the scrotum beyond the age of reversible changes take place in the testis. If the testis doe descend until puberty, it will be incapable of producing sperm thereafter. If the testes remain permanently within the abdomen after the age of 30 or 35, not only is spermatogenesis completely absent, but also hormonal secretion will be diminished.

The amount of connective tissue in the testis reflects its function. Elastic fibers appear in the tubular wall of patients with postpubertal hypogonadism, precocious puberty, or with arrested spermatogenesis. Various degrees of hyalinization of the tubular wall are associated with low androgen production. The intertubular connective tissue becomes more fibrous with regression of the seminiferous tubules in advanced age and in cases of testicular ischemia. Complete fibrosis of all or most of the tubules causes spermatogenic arrest.

Stress, anxiety and emotional tension have adverse effects on sperm production, and may alter hypothalamic function resulting in a failure of gonadotropin-releasing hormone (GnRH) to act on the pituitary. In most cases of paraplegia and quadriplegia, testicular atrophy is progressive. Denervation of the lumbar spinal nerves may affect both vascular tone and testicular temperature control.

Testicular biopsy could be used for the diagnosis of abnormalities of spermatogenesis, in the application of therapeutic procedures, and in the evaluation of the results of such therapy. Quantitative analysis of the biopsies permits a precise evaluation of the changes involving the germinal epithelium and/or Leydig cells.

Sperm and Semen Anomalies

It is difficult to define normal and abnormal sperm. Several criteria have been used to identify a normal and fertile ejaculate, and several descriptions have been used to define morphologically normal sperm. More is known about "sperm with normal morphology" than "normal sperm." Sperm with morphologically normal head, acrosome, midpiece and tail are more likely to be functionally normal with ability to penetrate cervical mucus and achieve fertilization of the egg.

Since sperm are normally discharged together with the prostatic fluid, disturbed prostatic function causes a decrease in the motility and survival of sperm. Different forms of sperm agglutination are encountered: head-to-head, tail-to-tail, or head-to-tail (Figure 50). Agglutinated sperm lose their directional motility. Aging is associated with a decline in fructose content of the seminal plasma and sperm count, and an increase in percentage of abnormal sperm.

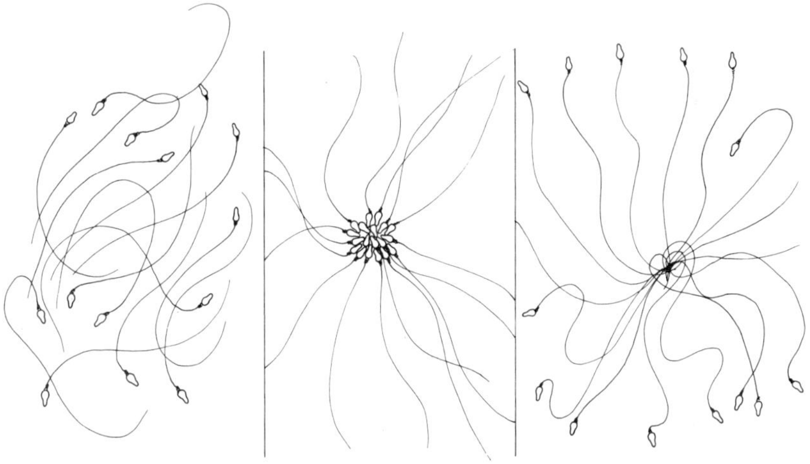

Figure 50. Different types of sperm agglutination interfering with fertilization of egg. Note head-to-head, and tail-to-tail, as compared to normal semen.

Ductal Obstruction

Azoospermia may be due to some mechanical block anywhere from the rete testis to the ejaculatory duct, postinfectious obliteration of the epididymal duct, or congenital malformation causing obstruction of the vas deferens or the epididymis.

Varicocele

Varicocele of the spermatic veins, usually unilateral, occurs in 8 to 20% of the healthy young adult men, 20 to 40 years of age. This syndrome is associated with decreased sperm count and lowered sperm motility. Testicular biopsies from infertile patients with varicocele show bilateral abnormalities characterized by hypoplasia of germinal cell with sloughing of immature forms into the lumen of the seminiferous tubules

Neurological Infertility

Sexual dysfunction can be associated with certain neurological disorders and multiple sclerosis. Sexual impotence may be due to endocrine or neurological disorders, or to anomalies in the external genitalia (see Chapter 10). Ejaculatory failure seems to be due to lack of contraction of the seminal vesicle, ampulla and vas deferens, whereas erectile disturbances may be due to disorders of libido, androgen deficiency, or neurological disorders. Failure of intromission and premature ejaculation are not uncommon.

Andrological Infections

Male fertility is greatly reduced by specific infection of the urogenital tract, *e.g.*, gonorrhea, syphilis, mycoplasma, trichomonas, *Candida albicans*, or tuberculosis. Gonorrhea causes abscess leading to testicular dysfunction. Certain cases of prostatitis and bacterial infection cause a decline in sperm motility. Semen with asymptomatic bacteriospermia show cytolytic and cytotoxic qualities, reduced sperm longevity and immunological response as agglutination. Asymptomatic infection of the genital tract is associated with sloughing of immature germinal cells from the seminiferous tubules.

Mycoplasma are spherical filamentous organisms that lack a cell wall but have a three-layer unit membrane, are gram-negative, contain DNA and RNA and are resistant to penicillin. T-mycoplasmas has been isolated in high frequencies from patients with various genital disorders. Since T-mycoplasmas may be attached to sperm, it may interfere with fertilization of the egg.

Several viruses are found in semen and prostatic secretions. Infectious cytomegalovirus (CMV) found in semen and urine may persist for several months. It has been isolated from semen of healthy young adults with and without history of mononucleosis and in patients in VD clinics. CMV has been implicated in spontaneous abortion and embryonic mortality. It is interesting to note that microbiological orgamisms contaminating semen can survive processing, freezing and storage, as well as can sperm cells. Various viruses and bacteria have been identified in liquid nitrogen, which is normally used for semen freezing.

Cytogenetic and Pathological Infertility

The most common cytogenetic disease is the Klinefelter's syndrome (Table XXII), which is characterized by small testes, azoospermia, elevated FSH levels and positive chromatin pattern. In a few cases (1.5%), infantile testes with elevated FSH are associated with normal chromatin patterns. In these cases, the patients have normal libido, but require androgenic support in later years.

Chronic renal failure is associated with endocrine dysfunction, impotence, testicular atrophy, alterations in testicular histology and gynecomastia. Hepatitis can also cause temporary, but often marked, depression of sperm production. Cystic fibrosis (disorder occurring in 1/1000 live births in North American Caucasian males) causes azoospermia when there is a congenital bilateral absence of the vas deferens.

REPRODUCTIVE FAILURE IN WOMEN

Reproductive failure in women may originate from physio-anatomical anomalies of reproductive organs, endocrine imbalance, menstrual disorders, chromosomal and congenital anomalies, gynecological infections, immunological and psychogenic factors, and anomalies of gestation.

Table XXI

Sex Chromosomes and Common Syndromes of Cytogenetic Infertility

Sex Chromosomes			*Total Number of*		
Sperm	*Egg*	*Offspring*	*Chromosomes*	*Syndrome*	*Clinical Features*
X	X	XX	46	Normal female	
Y	X	XY	46	Normal male	
XY	X	XXY	47	Klinefelter's[a]	Small testes, gynecomastia
XY	XX	XXXY	48	Klinefelter's[a]	
O	X	XO	45	Turner's[a]	Streak ovary, juvenile female genitalia, underdeveloped breasts, short stature
X	O	XO	45		
X	XX	XXX	47	Triple X[b] (superfemale)	Fairly normal female, secondary sexual characteristics may be poorly developed
Y	O	YO		Lethal	
O	O	OO		Nonviable	

[a] Klinefelter's and Turner's syndromes result from nondysjunction during oogenesis or spermiogenesis.
[b] Triple X is caused by faulty separation during oogenesis.

Physioanatomical Anomalies of Reproductive Organs

Structural anomalies in the size, shape and patency occur in the vagina, cervix, uterus, uterotubal junction and oviduct (Table XXIII; Figure 51).

Ovary

Various syndromes are associated with ovarian anomalies and endocrine imbalance (Table XXIII). Anovulatory cycles occur mainly in young girls and at the menopause, and also after childbirth. The follicle develops but does not reach ovulatory size. The corpus luteum does not develop and there is no change in basal body temperature.

Oviduct

Several disorders may adversely influence the structure and function of the oviduct, *e.g.*, (1) pelvic inflammatory disease (puerperal, Neisserian, tuberculos), (2) previous pelvic surgery, (3) endometriosis, and (4) tubal adhesions secondary to other processes.

Uterus

Anatomical abnormalities of the uterus involve its position in relation to the midline of the pelvis, and the direction of the axis of the cervix in relation to the axes of the vagina. Displacement of the uterus may be due to acute forward flexion caused by tumors. In genital prolapse, some reproductive organs protrude through an aperture between the two muscles and the orifice of the vagina. Prolapse may be due to obstetrical trauma, atrophy of the ligaments due to estrogen withdrawal, or congenital weakness of the ligaments.

Cervix

Reproductive failure may result from congenital anomalies of the cervix, mechanical factors, cervix-related surgical procedures, endocrine imbalance, abnormal cervical mucus or antigen-antibody reaction between semen and cervical secretions (Table

Table XXIII

Reproductive Failure as a Result of Anomalies of the Ovary and Female Reproductive Tract

Organ	*Pathophysiological Mechanisms*
Ovary	Failure to ovulate due to endocrine disorders, *e.g.*, disturbances of the pituitary or hypothalamus, or functional insufficiency of the pituitary
	Ovarian hypoplasia
	Cystic change: polycystic ovaries of the Stein-Leventhal syndrome, corpus luteum cyst, hemorrhagic cyst, theca lutein cyst
	Estrogen-secreting ovarian tumors
	Chronic infection
Oviduct	Failure of ovum transport in oviduct due to damage of mucosa, inflammation, occlusion, adhesions, tubal endometriosis, disturbances of tubal motility, or hypoplasia
Uterus	Congenital anomalies, failure of implantation of blastocyst due to atrophy of or inadequate progestational reaction in endometrium, irregular shedding of endometrium, hyperplasia of endometrium, myomas, inflammation such as endometriosis, or trauma
Cervix	Decreased surface area of cervical secretory epithelium as a result of surgery, traumatic scarring following induced abortion, postinfectious scarring, or developmental anomalies
	Cervical epithelium not responding to estrogen stimulation as a result of chronic endocervical infection
	Alteration in size (hypoplasia or hypertrophy) or displacement of cervix
	Endocrine imbalance, *e.g.*, exogenous progestins, estrogen inhibition
	Immunological factors causing agglutination or immobilization of sperm in cervical mucus
Vagina	Congenital vaginal atresia or aplasia
	Traumatic or inflammatory vaginal stenosis
	Infections: streptococci, staphylococci, trichomonas, fungus vaginismus

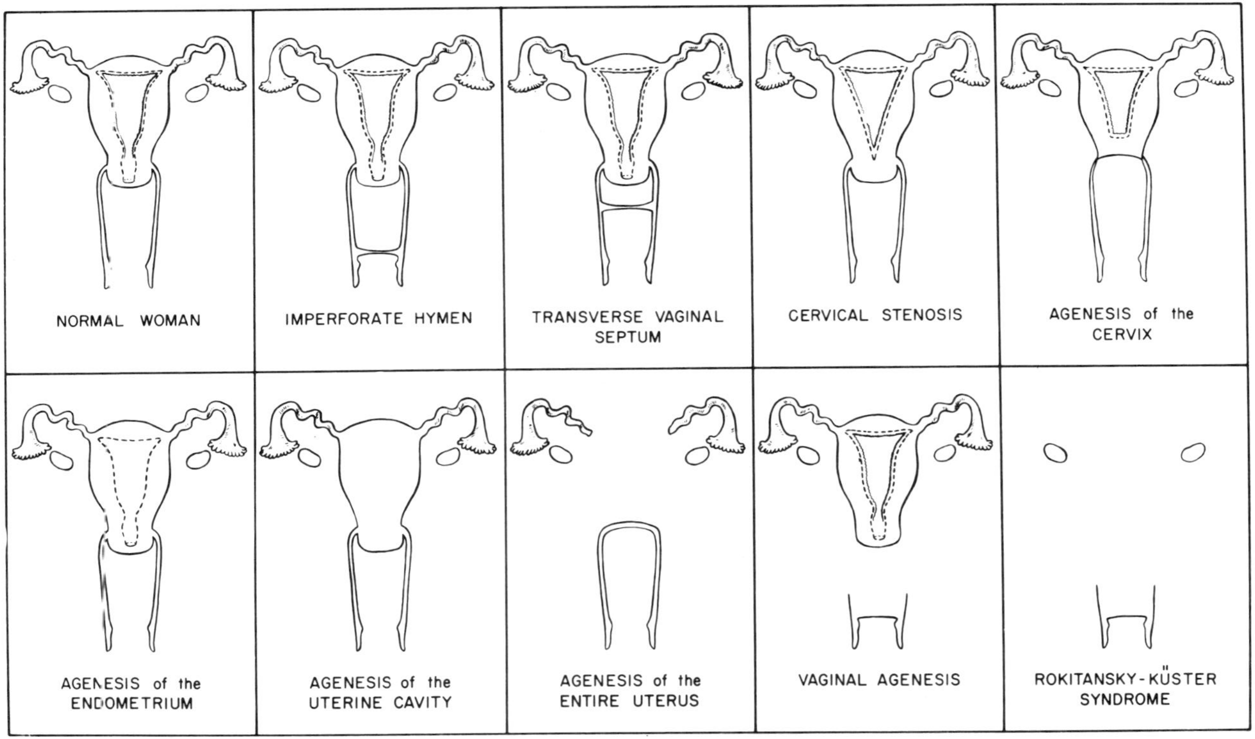

Figure 51. Developmental anomalies of the uterus and lower genital tract related to primary amenorrhea [From Somogyi and Jaszczak. In: *Sexual Maturity,* E. S. E. Hafez and J. Peluso, Eds. (Ann Arbor, Michigan: Ann Arbor Science Publishers, Inc., 1976).]

Table XXIV

Management of Cervical Mucus Factor to Correct Reproductive Failure

Characteristics of Cervical Mucus	*Management*
Deficient mucus	Treat ovarian dysfunction Administer estrogens during midcycle
Sperm cannot penetrate thick mucus	Treat ovarian dysfunction Administer estrogens during midcycle Treat cervicitis
Mucus leathal to sperm	Protected intercourse Artificial insemination (AIH)

XXIV). Congenital anomalies in the cervix, such as stricture of the cervix, imperfect canalization, or congenital atresia, may not necessarily cause reproductive failure.

Luminal Fluids

Idiopathic infertility may be due to disorders in the biochemistry in the luminal fluids of the female reproductive tract as a result of anomalies in cellular metabolism with subsequent loss of proper function of the secretory cells of the oviduct, uterus, or cervix. Several organic components of the luminal fluids play a major role in the survival of sperm, eggs and embryos during fertilization, cleavage and preimplantation. Except for the effect of pH on sperm survival, little is known concerning the effects of constituents of the vaginal fluids upon the motility, respiration and viability of sperm.

Vaginal discharge is derived from various sources of several components: vulval secretions from sebaceous, sweat, Bartholin's and Skene's glands; cervical mucus; endometrial and oviductal fluids; transudate through the vaginal walls; and exfoliated cells of the vaginal epithelium. Certain components of the vaginal fluid provide substrate for multiplication of vaginal microflora which, in turn, add their metabolic products to the secretions.

Anovulation

Hypothalamic GnRH reaches the anterior pituitary via the hypophyseal portal system and controls gonadotropic secretion. The next message in this molecular relay is conveyed by means of FSH and LH which, besides controlling their own secretions through a short loop feedback mechanism, also regulate ovarian function. Anovulation may be associated with luteinized unruptured follicle, galactorrhea, polycystic ovaries or other endocrine imbalance (Table XXV; Figure 52). Infertile women, with no apparent mechanical factors associated with their infertility, have been classified by Lunenfeld into seven groups, according to the etiology of the disturbance:

1. Hypothalamic-pituitary failure
2. Hypothalamic-pituitary dysfunction
3. Ovarian failure
4. Congenital or acquired genital tract disorder
5. Hyperprolactinemia with a space-occupying lesion in the hypothalamic-pituitary region
6. Hyperprolactinemia with no detectable, space-occupying lesion in the hypothalamic-pituitary region
7. Amenorrhea with a space-occupying lesion in hypothalamic-pituitary system

Luteinized Unruptured Follicle (LUF Syndrome)

Ovulation is clinically associated with BBT shifts, secretory endometrium and rise of serum progesterone. These criteria indicate luteinization of the follicle, with the presumptive inference of rupture of follicle and release of ovum. When laparoscopy was performed within a few days after presumed ovulation, inspection of the ovarian surfaces failed to show follicular rupture in some women. Rupture is defined by the presence of a stigma on vascularized corpus hemorrhagicum at that time of the cycle. The clinical significance of luteinized unruptured follicle in reproductive failure has not yet been evaluated.

Table XXV

Some Endocrine Anomalies Causing Reproductive Failure

Endocrine Anomaly	*Syndrome*	*Ovarian Anomaly*	*Clinical Characteristics*
Excess of Estrogen	Precocious puberty	Follicular cyst Polycystic ovary	Menorrhagia, amenorrhea, enlarged breasts, secondary sexual development
	Hyper- and polymenorrhea	Granulosa theca cell Tumor	Endometrial hyperplasia
	Postmenopausal bleeding	Sertoli-Leydig cell tumor	Cornified vaginal mucosa
Deficiency of Estrogen	Defeminization (failure of feminization)	Ovarian agenesis (atrophy) Polycystic ovary; masculinizing ovarian tumor	Oligomenorrhea, amenorrhea, regression of secondary sex characteristics, atrophy of genital organs and breast
Excess of Androgen	Masculinization (with or without defeminization)	Polycystic ovary, Sertoli-Leydig cell tumor	Hirsutism, enlarged clitoris, temporal hair recession, baldness, enlarged larynx, masculine habitus
Excess of Corticosteroid	Cushing's Syndrome (with or without masculinization)	Lipoid cell tumor	Adiposity of trunk, hypertension, diabetic tendency, osteoporosis

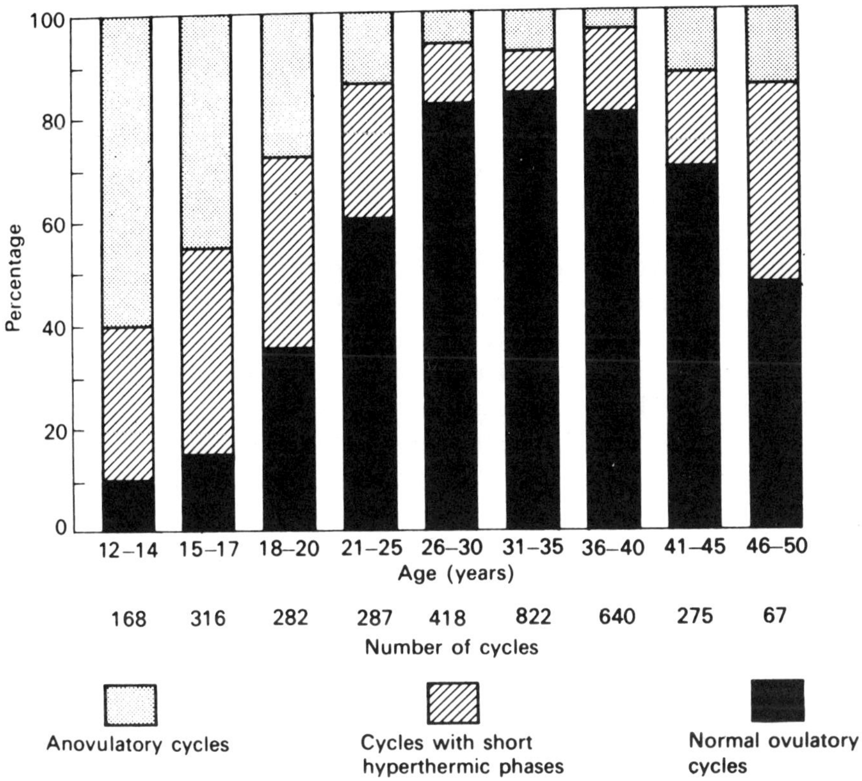

Figure 52. Anovulatory, incomplete and normal menstrual cycles in girls and women. Both types of atypical cycles are more frequent in girls and older women, whereas normal cycles are most frequent at 25-35 years of age. (From Doring, G. K. "Uber die relative Haufigkeit des anovulatorischen cyclus in Leben," *Arch. Gynaek.* 199:155, 1963.)

Galactorrhea and Ovarian Dysfunction

Although galactorrhea and hyperprolactinemia are often used synonymously, galactorrhea can occur in the presence of normal prolactin levels. On the other hand, hyperprolactinemia without lactation is not uncommon. Low concentrations of prolactins are required for gonadal steroidogenesis. However, high levels of prolactin may block the action of gonadotropin on gonadal function and interfere with steroid secretion. Thus, varying degrees of ovarian dysfunction, from abnormal menstrual cycles to amenorrhea, may occur. Most patients with hyperprolactinemia have hypogonadism which occurs irrespective of the cause of elevated prolactin levels. Administration of bromergocryptine or L-Dopa suppresses prolactin secretion, inhibits galactorrhea and restores cyclic ovarian function.

Polycystic Ovary

The polycystic ovary, regardless of its size, is an expression of a multiglandular disorder of the adrenal, hypothalamus and/or the ovary. The immediate consequence of a polycystic ovary is anovulation. The treatment consists either in ovulation induction or in suppression of ovarian function with the "pill" and corticoids, if needed. The treatment of the usual polycystic ovary is similar to that of anovulation and of hyperandrogenism, after a careful differential diagnosis to rule out undetected tumors which may cause similar symptomatology.

Corpus Luteum Defects

Menstrual cycles with corpus luteum defects are characterized by shortened luteal phase caused by decreased corpus luteum activity leading to decreased ovarian steroid production, and retarded development of the secretory endometrium with subsequent inhibition of implantation. Corpus luteum defect is primarily due to inadequate or inappropriate gonadotropic stimulation as a result of inadequate follicular maturation in the ovary; inadequate LH surge; increased luteolytic activity of the corpus luteum; and/or disturbed endometrial response. Diagnosis of luteal phase defect is based on short luteal phase (less than 10 days) as judged by BBT; dated endometrial biopsy; and low serum progesterone during the second half of the cycle.

Insufficient follicular ripening characterized by lowered serum estradiol during prolonged follicular phase leads to defective Graafian follicle and subsequent defective corpus luteum. Defective follicular ripening may be due to hypostimulation of the ovary or to inadequate follicular system. Luteal phase defects occurring after normal follicular ripening and characterized by normal serum estradiol are probably due to an inadequate LH surge and low LH during luteal phase.

Luteal phase defects with normal or high estradiol, high testosterone and low progesterone occur in gonadotropin or GnRH-stimulated cycles of women with polycystic ovary syndrome. Polycystic ovaries during the luteal phase produce more testosterone and estradiol at the expense of progesterone. Replacement therapy with progestins is only recommended in minor forms of luteal insufficiency. In cycles with a luteal phase shorter than nine days, treatment with clomiphene or gonadotropins is recommended.

Menstrual Disorders

Disorders of menstruation include amenorrhea (absence of menstruation), dysmenorrhea (painful menstruation), or menorrhagia (excessive menstruation).

Amenorrhea

Primary amenorrhea, delayed menstruation by 18 years of age, may be due to endocrine or developmental disorder (Figures 53 and 54). In secondary amenorrhea, menstruation ceases after it has been established for months or years. Physiological amenorrhea occurs during pregnancy, breastfeeding and postmenopausal life. Pathologic amenorrhea, however, results from one of the following disturbances (Table XXVI):

1. developmental (obstructive)–vaginal or cervical, or imperforate hymen;
2. hypothalamo-pituitary-ovarian axis–pseudocyesis, anorexia nervosa, pituitary adenoma, or ovarian dysgenesis;
3. adrenal hyperplasia, Addison's disease, hyperthyroidism, or hypothyroidism (usually menorrhagia); or
4. uterine atrophy or aplasia, or Asherman's syndrome (absence or defective endometrium).

Post-Pill Amenorrhea

Sex steroids and several drugs interfere with the synthesis or release of hypothalamic GnRH. Estrogens and progestogens also act directly on the pituitary gland and gonadotropin release.

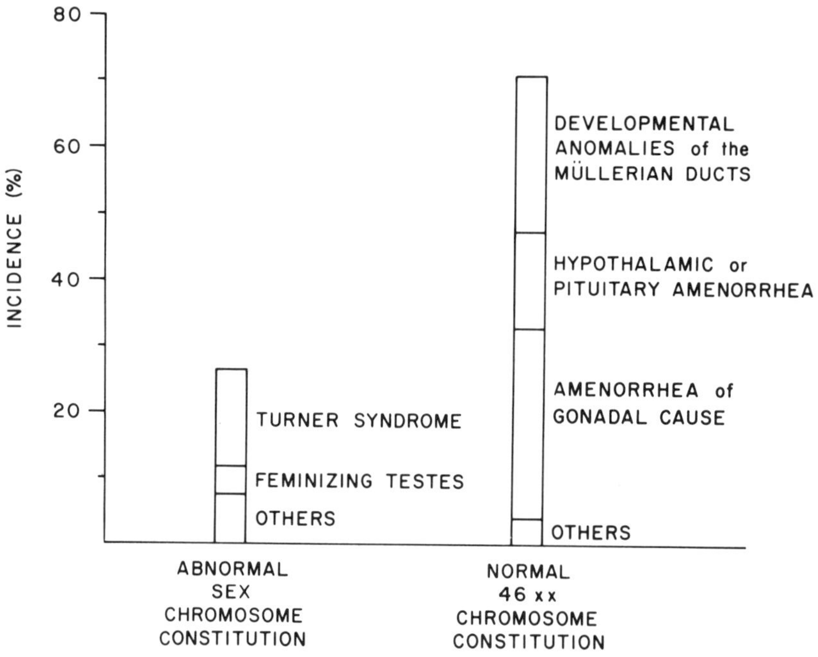

Figure 53. Incidence of the main syndromes associated with primary amenorrhea. [From Somogyi and Jaszczak. In: *Sexual Maturity*, E. S. E. Hafez and J. Peluso, Eds. (Ann Arbor, Michigan: Ann Arbor Science Publishers, Inc., 1976).]

Suppression of hypothalamic-pituitary axis and anovulation are the principal methods by which steroidal contraceptives exert their antifertility effect. Following the administration of oral contraceptives, persistent anovulation and amenorrhea may be associated with galactorrhea in certain women. Functional recovery of hypothalamic-pituitary-ovarian axis after termination of oral contraceptive therapy is usually prompt.

Amenorrhea after oral contraceptive therapy is a self-limited condition; in more than 90% of patients, spontaneous menstruation resumes within 6 to 12 months. Clinical management requires an investigation of hypothalamic, pituitary, ovarian, adrenal and occasionally thyroid disorders.

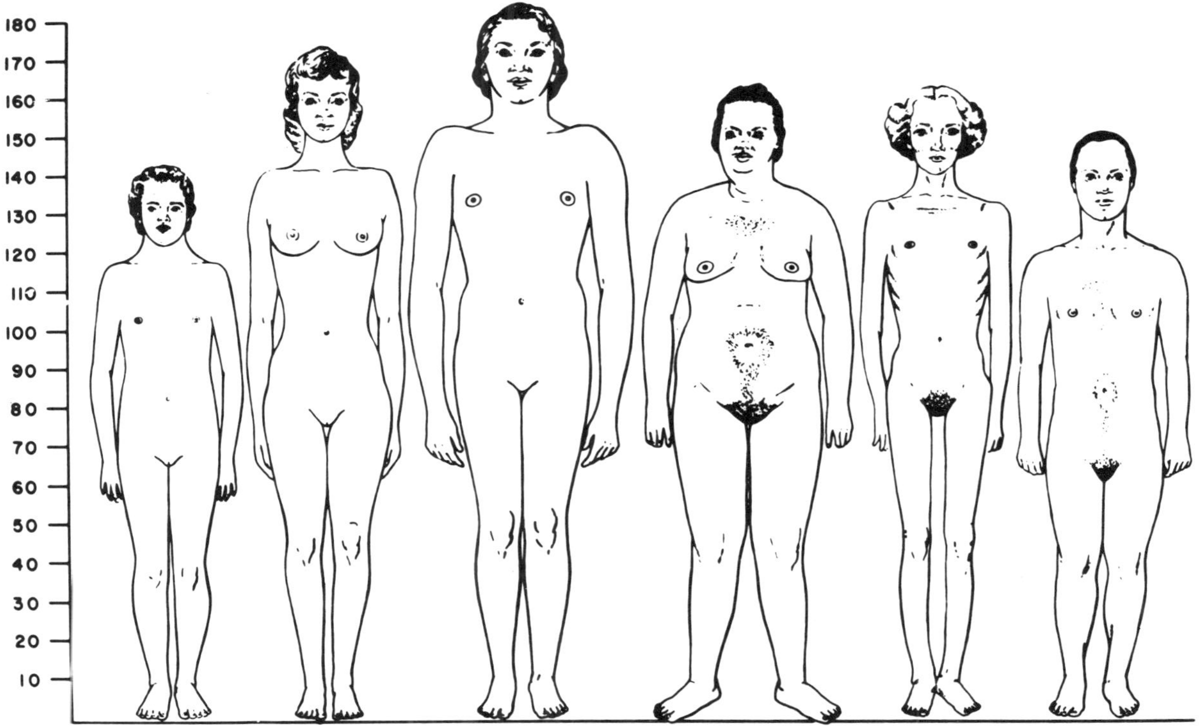

Figure 54. Somatic features and primary amenorrhea: Turner's Syndrome; feminizing testes; gonadal dysgenesis; Stein-Leventhal Syndrome; anorexia nervosa; adrenogenital syndrome. [From Somogyi and Jaszczak. In: *Sexual Maturity,* E. S. E. Hafez and J. Peluso, Eds. (Ann Arbor, Michigan: Ann Arbor Science Publishers, Inc., 1976).]

Table XXVI

Types of Amenorrhea

Type	*Clinical Symptoms*	*Physiological Causes*
Physiological Amenorrhea	Absence of menstrual cycles during pregnancy, lactation, breast feeding, postmenopause	Suppression of hypothalamo-pituitary-ovarian functions
Primary Amenorrhea	Delayed menarche, delayed onset of secondary sexual characteristics after age 16	Ovarian malfunction; other endocrine disturbances
Secondary Amenorrhea	Absence of menstruation after normal menstrual functions	Disturbed endometrium: congenital (Asherman's syndrome) irradiation Ovarian anomalies: premature ovarian failure, or castration due to surgery or irradiation Anomalies in pituitary-hypothalamus axis Nutrition: starvation or anorexia nervosa Tumors: masculinizing ovarian tumor, feminizing ovarian tumor, neoplasia Infections
Post-Pill Amenorrhea	Oversuppression as a result of combined estrogen-progestin pills, may persist for prolonged periods with spontaneous recovery in some patients	Suppression of hypothalamic-pituitary axis by exogenous steroids
Pseudycyesis	Amenorrhea associated with morning sickness, abdominal enlargement, breast changes, possible perception of fetal movements	Psychogenic

Dysmenorrhea

Dysmenorrhea is associated with ovulation and the development of the corpus luteum. If ovulation is prevented by the administration of hormones, dysmenorrhea does not occur. Pubertal menstrual cycles are usually anovulatory. The pain associated with dysmenorrhea is probably caused by ischemia due to the vigorous contractions of myometrial muscles during the first days of menstruation. Primary dysmenorrhea may be due to obstruction to the flow from the uterus by a narrow cervical os or malposition of the uterus; intensive uterine contractions before and during menstruation; or vasoconstriction of the terminal arterioles before the shedding of the endometrium.

Menorrhagia

Menorrhagia refers to frequent menstrual periods and/or excessive menstrual blood loss. Menorrhagia, a symptom not a disease, may be caused by fibroids increasing the area of the uterine cavity; inflammation causing endometritis or salpingitis; certain blood disorders; thyroid dysfunction such as myxoedema; or estrogen-progestogen imbalance. Excessive uterine bleeding may result from dysfunction uterine bleeding, infection, or trauma. Adolescent bleeding (pubertal menorrhagia) is characterized by prolonged episodes of bleeding, anemia, anovulatory cycles and endometrial hyperplasia. Postmenopausal bleeding, which occurs one year or more after menopause, may indicate the presence of genital malignancy.

Chromosomal Anomalies and Congenital Anomalies

Chromosomal anomalies cause intersexuality and reproductive failure (Table XXII; Figure 54). Turner's Syndrome (gonadal dysgenesis) causes delayed pubertal growth and development. Monosomy of the X-chromosome or its short arm and mosaicism of sex chromosomes are associated with sexual infantilism, "streak gonads," absent adolescent growth spurt and short stature.

Birth defects may be due to hereditary factors, mutations, chromosomal aberrations, maternal disease, drugs, infection of the uterus, placenta or fetus, or insufficiency of placental blood supply. Maternal respiratory conditions that seriously impair oxygen supply to the mother will also affect oxygen levels. About 20% of mothers of diplegic infants suffer from physical disorders, chronic infections, anemia, cardiac disease or gynecological disorders.

The uterus, oviducts and upper part of the vagina develop from the two Mullerian ducts which fuse to form the single uterus and vagina (Figure 51). When these ducts fail to fuse, they give rise to various congenital anomalies. In extreme cases, there are two uteri and two vaginae.

Malformations, which are morphological changes originating during prenatal life, may affect one or more organs, one or more systems, or the entire body. Congenital abnormalities occur more frequently in monozygotic than in dizygotic twins, probably because monozygosity in itself is an anomaly. Not all congenital anomalies, however, are due to genes; many occur from accidents during development and therefore are not transmitted to further progeny. Common phenotypic expressions of hereditary origin are due to absence of gonadal tissue, production of abnormal gametes, or the production of fertilizable ova, which result in inherently defective embryos.

Gynecological Infections

Gonococci, staphylococci, streptococci, pneumococci and coliforms may cause inflammation of the oviduct, uterus, cervix, or vagina. Infection with T-strain mycoplasma may be a potential factor in sperm immobilization in cervical mucus or early abortion. Subclinical pathologies by infections may affect sperm transport to the upper portions of the reproductive tract by potentiating the immune response against sperm.

Immunological and Psychogenic Factors

Precise knowledge of immunological and psychogenic factors is lacking. Women may develop antibodies against sperm. These antibodies can be demonstrated within the serum and cervical secretion. Sperm-immobilizing or -agglutinating antibodies may function at the level of the cervix or uterus, and may not be detectable on postcoital examination or even on serologic evaluation of peripheral blood.

Psychogenic factors of reproductive failure include emotional conflicts, frigidity or vaginismus, unstable marriage, poor adjustment to the sexual partner and environmental factors. These psychogenic factors may be deeply rooted and require expert psychiatric help, or may be due to lack of understanding of human sexuality. Psychoneuroendocrinological infertility, associated with functional hypothalamic amenorrhea, includes anorexia nervosa, pseudocyesis, persistent corpus luteum syndrome, and subtle psychological trauma (usually at time of puberty).

Anomalies of Gestation

Anomalies of pregnancy include prenatal and perinatal mortality, ectopic pregnancy and abortion.

Prenatal Mortality

The failure of the mammalian blastocyst to implant may be a result of maternal-fetal vascular deficiencies or biochemical changes in the endometrium, endometrial secretions or trophoblast. Prenatal mortality, which is more frequent than previously realized, may occur at all stages of gestation.

Ectopic Pregnancy

In ectopic pregnancies the blastocyst does not implant in the uterine cavity. Some 95% of ectopic pregnancies occur in the oviduct and the rest occur in the uterine cornu, ovary, cervix, abdominal cavity, or broad ligament (Figure 55). Tubal pregnancy may be due to infections or congenital malformations of the oviduct associated with destruction of tubal cialia, presence of patches of endometrium in the oviduct, delayed ovum transport in the oviduct, or fibrosis following salpingitis.

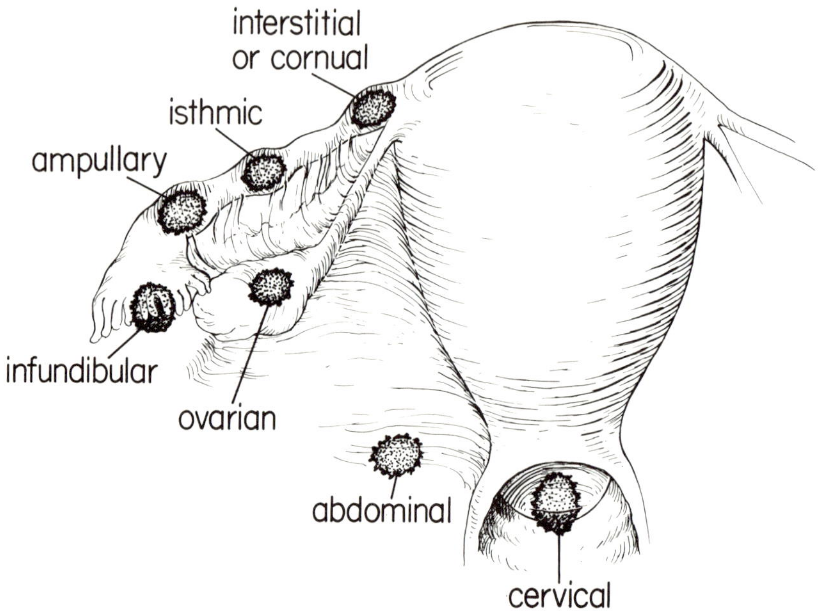

Figure 55. Sites of ectopic pregnancy. Abdominal ectopic pregnancy may occur in the broad ligament and adjacent visceral and parietal surfaces.

Abortion

Abortion may occur at any time before the 28th week of pregnancy but is most common at 8 to 12 weeks (15% of all pregnancies), particularly in cases of spontaneous or hormone-induced multiple pregnancy. Abortion is attributed to fetal, maternal or genetic factors, such as congenital malformation and abnormalities of the chorion, faulty implantation in unfavorable uterine site, uterine fibroids, congenital malformation or retroversion of the uterus, incompetence of the internal os of the cervix, and progesterone deficiency during pregnancy. Abortion may also result from trauma, emotional shock, acute febrile illness, septicaemia with infection of the fetus or placenta, untreated syphilis, severe hypertension, renal disorders, viral infections of pathological origin, early pregnancy following puberty, nutritional deficiencies, or certain drugs such as ergot, quinine and lead.

Table XXVII

Some Laboratory Tests Used for Diagnosis of Female Infertilit

Laboratory Test		*Indications*
Endocrine Profile (RIA)	Gonadotropins: LH, FSH, prolactin	Endocrine imbalance Infertility Ovulation induction Inadequate corpus luteum
	Estrogen Pregnanediol	Endocrine imbalance Infertility
	Androgen	Hirsutism Virilization Intersexuality
Endocrine Function	Progestin test Gonadotropin test Dexamethasone test	Secondary amenorrhea Ovulation induction Hirsutism Virulization
Karyotyping (chromosome analysis)	Blood culture Tissue culture	Primary amenorrhea Turner's syndrome
Sex chromatin	Smear from buccal or vaginal mucosa	Primary amenorrhea
	Drumsticks in blood smear	Intersexuality

DIAGNOSTIC PROCEDURES OF INFERTILITY

Reproductive failure of the "couple" as a unit should be evaluated. The term "infertile" versus "sterile" of a couple should be recognized in specific physiological terms. Several procedures are used for the diagnosis of infertility, *e.g.*, clinical examination, analysis of endocrine profile, monitoring of ovarian function, detection of tubal patency, *in vivo* and *in vitro*

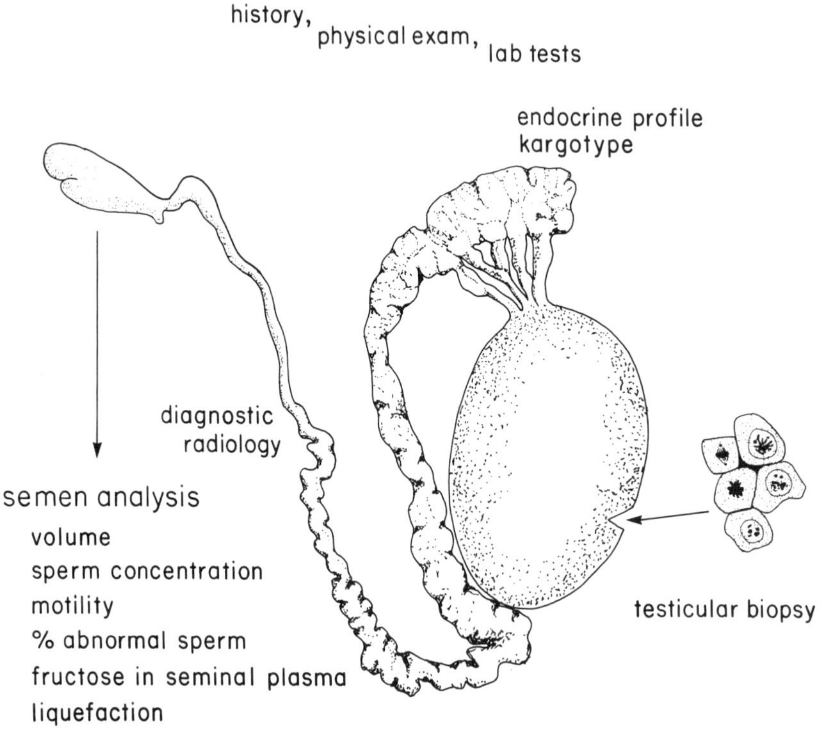

Figure 56. Various clinical and laboratory techniques for diagnosis of reproductive failure in men.

postcoital tests, biochemical analysis of vaginal discharges and semen analysis (Table XXVII; Figures 56 and 57).

Medical History and Clinical Examination

Medical history and clinical examination of infertile men may reveal anomalies in the size, consistency of the testes, epididymis, vas deferens and prostate (Figure 58). Testing of voluntary and reflex function of the pelvic floor muscles are helpful diagnostic tools to determine neurological infertility.

Medical history of infertile women includes menstrual history, *e.g.*, age at onset, type of cycle, duration of flow and dysmenorrhea. Clinical examination includes character, amount of vaginal discharges, general physical development, evidence of endocrine disorder, or related diseases. Abdominal examination for

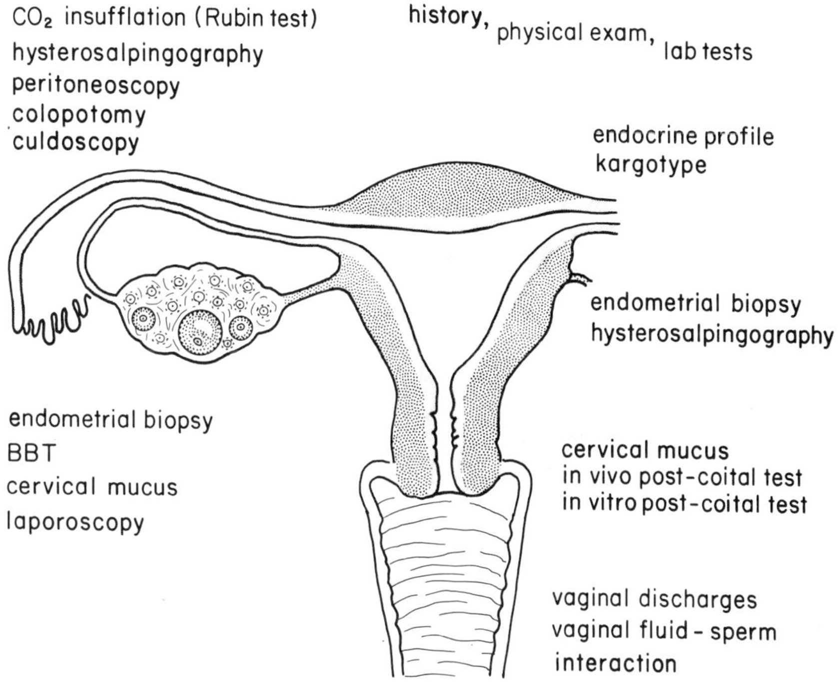

Figure 57. Various clinical and laboratory techniques for diagnosis of reproductive failure in women.

scars and masses is followed by vaginal examination. The appendages are palpated for any enlargement or thickening of the tubes or ovaries. A speculum is used to examine the condition of cervix and type of cervical secretion in relation to time in menstrual cycle.

Diagnosis of Endocrine Disorders Using Radioimmunoassay

Biological methods of hormone assay were cumbersome and complicated. Quantification of gonadotropins and steroids is easily performed by radioimmunoassay (RIA). RIA utilizes the specificity of the antigen-antibody reaction in combination with the sensitivity and semiautomated features of radiochemistry. Thus, picograms of steroids or nanograms of gonadotropins can

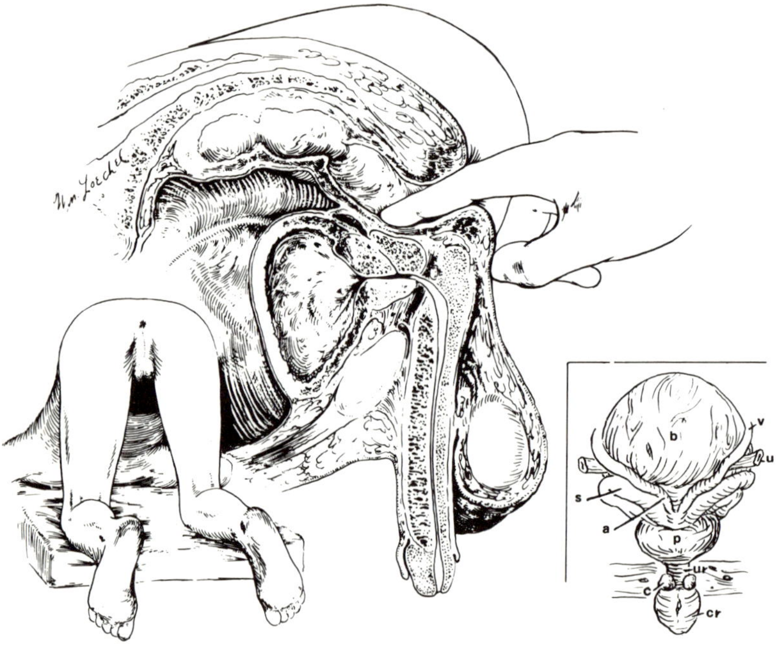

Figure 58. Clinical examination of the prostate. The patient is placed in knee-chest position. Prostate is massaged through the rectum from its proximal to its distal area. The inset shows details of (a) ampulla; (b) bladder; (c) Cowper's gland; (cc) corpus cavernosum; (s) seminal vesicle; (u) ureter; (ur) urethra; and (v) vas deferens. [From Pontes and Pierce. In: *Techniques of Human Andrology*, E. S. E. Hafez, Ed. (Amsterdam: Elsevier, 1977).]

be measured accurately in a few tenths of one ml of the patient's serum (Table XXVIII).

The concept of radioimmunoassay is based on the competition between two reactions occurring simultaneously in an assay tube as follows:

Antibody + "Radiolabeled" hormone   "radioactive" immunocomplex
Antibody + "cold" hormone   "nonradioactive" immunocomplex

When the concentration of antibody and "radiolabeled" hormone is constant, the amount of radioactivity in the immunocomplex is inversely related to the amount of "cold" hormone

Table XXVIII

Normal Laboratory Values of Serum Hormones of Reproduction in Fertile Women (From Maghissi. Unpublished data and the literature)

Hormone			*Values*
FSH (mIU)			3-50
LH (mIU)			6-100
Estrone (E_1)	(pg/ml)	follicular	25-50
		midcycle	45-300
		luteal	45-200
Estradiol (E_2)	(pg/ml)	follicular	20-200
		midcycle	150-400
		luteal	120-350
Progesterone	(ng/ml)	ovulatory	3-20
		anovulatory	0-0.3

in the sample. With the addition of excess labeled hormone to plasma samples, all binding sites are occupied. In the absence of unlabeled hormone, all binding sites are bound to radioactive hormone. In the presence of unlabeled hormone, it competes with radioactive hormones for binding sites and less label is taken up. In the presence of an excess of unlabeled hormone, few sites (essentially none) can be occupied by radioactive hormone. A curve is produced by plotting known quantities of unlabeled hormone against the amount of label bound. Test samples are then interpolated to assay hormone levels in the plasma. By the use of a standard dose response curve, the amount of gonadotropin or steroid in the tested sample can be computed. RIA procedures are best accomplished by computer-processing all data, using built-in quality control features.

Endocrine disorders in women can be detected by estimation of FSH, LH, estrogen, progesterone and androgens in blood plasma. Endocrine disorders in men can be detected by estimation of fructose levels in the semen or plasma gonadotropins and steroids. The administration of synthetic GnRH is used to differentiate pituitary from hypothalamic disorders (Table XXIX).

Table XXIX

Synthetic-Releasing Hormone Employed in Tests of Pituitary or Hypothalamic Dysfunctions

Disease Tested	*TRH Level*	*GnRH Level*
Pituitary tumor or destruction of gonad	Absent to normal	Absent to normal
End organ failure:		
thyroid	Exaggerated	
ovary		Exaggerated
Idiopathic, single or multiple failure	Normal	Low to normal Rarely any

Monitoring Ovarian Function

Normal functions of the ovaries can be monitored by histological examination of the endometrium, endocrine profiles, BBT and examination of biophysical characteristics of cervical mucus. Histological examination of the endometrium following biopsy or a dilation and curettage (D&C) during the premenstrual phase of the cycle will produce evidence of the secretory phase if ovulation has occurred. Tuberculosis is excluded by histological examination and by culture and guinea-pig inoculation. A daily record of basal body temperature (BBT) is kept for a period of three months. Cyclical variations in BBT may indicate the most fertile days since a slight rise in daily temperature occurs after ovulation.

Tubal Patency

Several diagnostic procedures are used to evaluate tubal patency: CO_2 insufflation, hysterosalpingography and peritoneoscopy with concomitant oviductal lavage. None of these procedures enables interpretation of the dynamics of oviductal contractility or provides information regarding characteristics of oviductal secretions.

Carbon dioxide, insufflated into the uterine cavity just after menstruation, leaks out of the tubes into the peritoneal cavity causing a fall in the manometeric pressure. Using a stethoscope, gas may be heard bubbling into the abdomen.

Tubal patency can be visualized through the laparoscope. After induction of a pneumo-peritoneum with CO_2, the laparoscope is introduced to inspect the uterus, oviducts and ovaries. A cannula is inserted into the cervix to inject a solution of methylene blue. In patent tubes, the dye passes through the infundibulum to accumulate in the pouch of Douglas. Tubal obstruction is easily recognized.

Hysterosalpingography is done where gas does not pass, and in cases of salpingitis, or malformations of the uterus. A radio-opaque dye (solution of iodipamide methylglucamine or sodium diatrizoate) is injected and its passage is observed on the X-ray screen or X-ray film.

Postcoital Tests

In Vivo *Postcoital Test*

At midcycle near the time of ovulation, cervical mucus is aspirated as early as 2 to 8 hours after intercourse or as late as 24 hours postcoitum. The mucus is examined microscopically to count the number of motile sperm per high-powered field (HPF) (Table XXX; Figure 59). Several variables must be considered in the interpretation of the results:

1. interval of abstinence before the test;
2. day of the cycle with respect to ovulation on which the test was made;
3. time interval between intercourse and examination of the mucus;
4. quantity and biophysical characteristics of cervical mucus;
5. adequacy of coital technique; and
6. quality of the ejaculate.

In vivo postcoital tests may reveal faulty coital technique, excessive acidity from chronic cervicitis inhibiting normal sperm activity, and poor quality cervical mucus.

Table XXX

The Postcoital Test (PCT) as a Diagnostic Tool of Infertility

Period of abstinence before PCT	2 days	
When to perform PCT	2 days before BBT rise; 2-6 hr after intercourse	
Spinnbarkeit of cervical mucus at PCT	6 cm	
Number of sperm per high power field (X 400)	>5	Good
	Only dead sperm	Poor
	No sperm	Negative

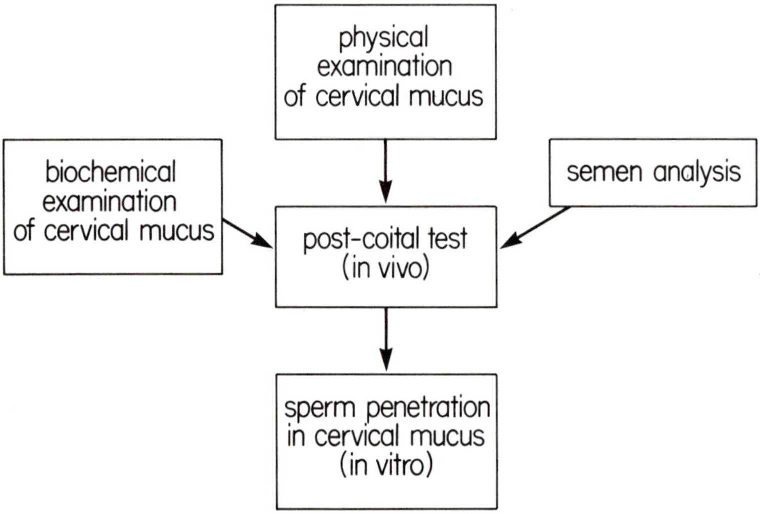

Figure 59. Evaluation of cervical factor in reproductive failure.

In Vitro *Postcoital Test*

When a drop of semen comes in contact with normal cervical mucus on a microscopic slide, a sharp interface forms. From this junction, a finger-like projection of semen protrudes into the mucus and sperm penetrates this wedge, progressing into the mucus at a rate of up to 3 mm/min. An *in vitro* test is carried out with cervical mucus collected in a catheter. The catheter containing the mucus is placed vertically in a glass cup containing a fresh semen specimen obtained from the patient's husband (Figure 60). After 90 minutes of incubation, the column of mucus is partitioned into four segments and examined as the *in vivo* test. The interpretation of results is simple when a large number of motile, normal-appearing sperm (10 to 20/HPF or more) are found in the mucus

Vaginal Discharge

The intensity and pleasantness of vaginal odors change throughout the menstrual cycle. Little is known about the possible effect of vaginal odors on psychosexual patterns of men. Vaginal discharges contain a variety of organic metabolites from different parts of the reproductive tract. Qualitative studies of vaginal fluids were done with gas chromatography and gas chromatography-mass spectrometry for the separation and identification of the constituents. Abnormal metabolites or anomalous concentrations of known compounds in the metabolic profiles provide a diagnostic marker for particular pathologies.

Semen Analysis

Semen analysis is used to evaluate the function of seminiferous tubules. A masturbation specimen is best suited for semen analysis, which should include ejaculate volume; sperm concentration; sperm motility or forward progression; and percent of sperm abnormality. A normal ejaculate has a volume of 2 to 3 ml with a concentration of 200 million sperm/ml (Table XXX). A concentration lower than 40 million/ml and with a larger number of abnormal sperm is indicative of infertility. At least 40% of the sperm should exhibit normal progressive (forward) motility.

THERAPEUTIC APPROACHES

Several hormones and drugs have been used with varying degrees of success to treat male infertility such as adrenergic drugs, cortisone, caffeine, theophylline, arginine, zinc, vitamins, ascorbic acid, clomiphene, HCG, HMG and GnRH. Artificial

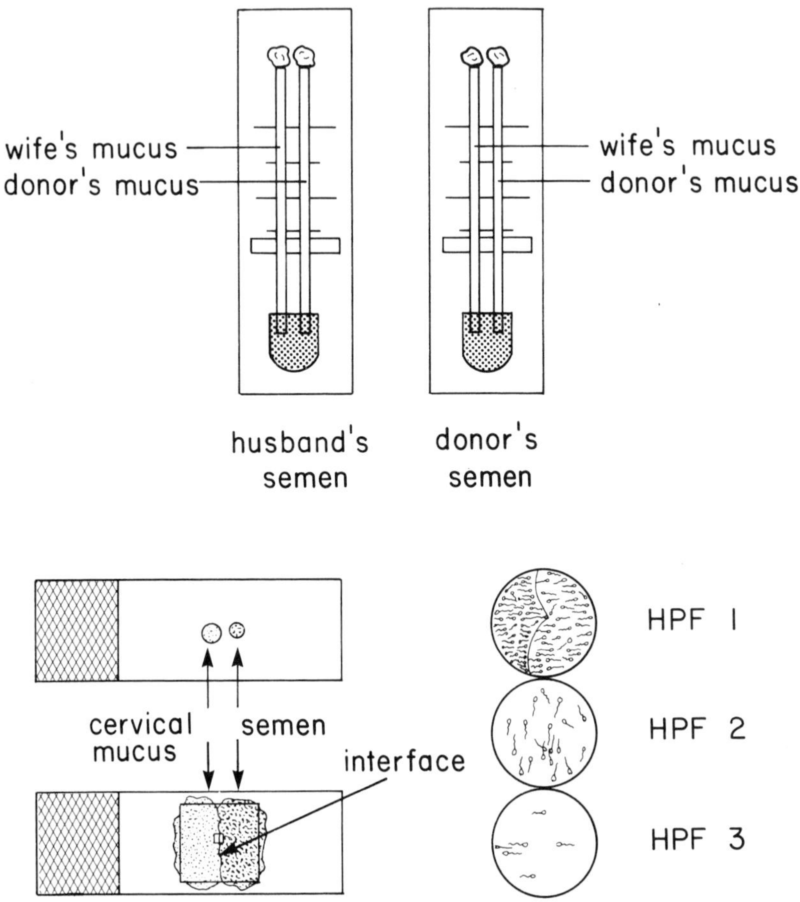

Figure 60. *In vitro* postcoital test (PCT) to evaluate penetrability of sperm in cervical mucus:

(a) Capillary method of Kremer: cervical mucus collected from the wife and a donor is placed in a glass cup containing husband's and donor's semen. The degree of penetration of sperm is evaluated at intervals.

(b) The slide test: a coverslip is placed on two droplets of semen and cervical mucus. The number of sperm penetrating the mucus are counted at high power at three adjacent microscopic high-power fields (HPF_1, HPF_2 and HPF_3).

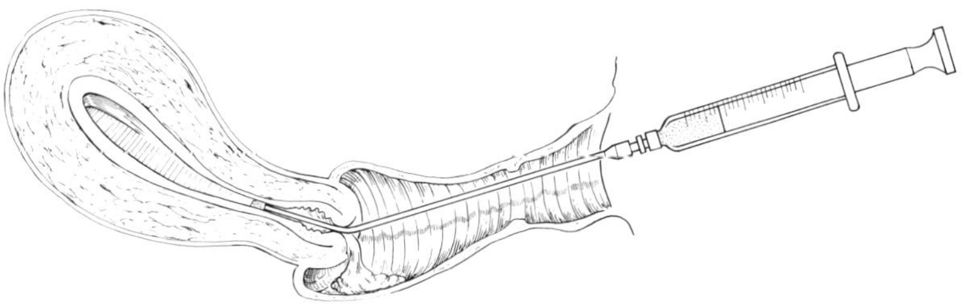

Figure 61. Intracervical artificial insemination with husband's (AIH) or donor's (AID) semen.

insemination with husband's semen (AIH) or a donor's semen (AID) have been used in certain cases of reproductive failure (Figure 61). Abnormal liquefaction of the ejaculate constitutes a frequent problem in male infertility. Amylase, kallikerin and Alivaire (a mucolytic agent) have been used with varying degrees of success in improving the fluidity of these abnormal ejaculates. Surgical approaches to male sterility include varicocelectomy (removal of varicocele), vasoepididymal anastomosis and vaso-vasostomy.

Several hormones and drugs have been employed to induce ovulation in anovulatory women, *e.g.*, estrogen, corticosteroids, clomiphene citrate, ergocryptine, GnRH, HMG-HCG and HPG-HCG. Clomiphene citrate, HCG and progesterone during the second half of the cycle have been used to treat defective luteal phase. Various surgical approaches have been used to correct female infertility.

Embryo transfer, extensively used for laboratory and farm animals, has been attempted in man. *In vitro* cleavage of human eggs with subsequent blastocyst development has been reported. It is possible that the techniques of *in vitro* fertilization and egg transfer will become common in gynecological practice. The ideal couple for this technique would be a fertile husband whose wife has blocked oviducts but an otherwise normal reproductive tract. Transplantation of ovaries with successful birth of young has been reported in rabbits. With better understanding of immunological problems associated with organ transplantation, it may be possible to correct certain reproductive failures by transplanting ovaries, oviducts and/or uteri.

CHAPTER 12

REPRODUCTIVE SENESCENCE

In man, most nonhuman primates, and some small laboratory rodents, reproductive ability ceases well before the end of the normal life span. Females of other mammalian species produce young, although with increasing difficulty, until late in life. Aging is associated with dramatic morphological and physiological changes in the reproductive organs, hypothalamus, pituitary gland and other endocrine glands. Little is known about the psychological, social and cultural aspects of reproductive behavior among the aged because of taboos on sexuality and the incorrect belief that the aged have no sexual needs. Sexual behavior of the aged, however, is influenced by physical illness, emotional disturbances and cognitive impairment.

MENOPAUSE

Menopause, the final phase of the reproductive ability of the woman, refers to the cessation of menstrual bleeding. "Climacteric" is used more correctly to include the wide range of symptoms occurring at this time in a woman's life. The average age of onset of menopause is 50 years (range: 37 to 56 years) and seems to be steadily increasing in the highly developed societies. Thus, the average length of the menstrual life is 36 years (range: 22 to 45 years). Menopause is completed after one year of amenorrhea, normally by 55 years of age.

The manner in which menstruation ceases is variable. Occasionally, it may cease abruptly, but as a rule the menstrual flow becomes gradually more scanty, and the interval between menstrual periods more prolonged (Figure 62). Failure of ovulation causes missed or shorter and less profuse periods, or functional uterine bleeding with prolonged and grossly irregular cycles. Occasionally, menstrual bleeding made of black or dark brown scanty flow, becomes less frequent until it ceases. Irregular menstrual cycle length may be due to decreased sensitivity of the hypothalamus to the feedback control of ovarian estrogen.

Premature menopause is caused by ovariectomy, irradiation of the ovaries, debilitating disease, or endocrine imbalance associated with decreased estrogen. Induced menopause frequently leads to shorter, but more severe, menopausal symptoms, whereas delayed menopause may be due to carcinoma or ovarian tumors.

Clinical Symptoms

Menopause is associated with several disturbances. Vasomotor instability causes irregularity in the diameter of peripheral blood vessels. Such fluctuations permit more blood to flow at times, causing flushing of the skin, a sensation of warmth and profuse sweating. Hot flushes are due to the high levels of gonadotropin. Estrogen deficiency may also be involved since estrogen therapy depresses gonadotropin secretion and controls the flushes. Other common symptoms include fatigue, dizziness, migraine headaches, chest and neck pains, excessive desire for sleep, nervous tension and irritability. Facial features undergo changes, wrinkles grow deeper and bags form under the eyes. There may be a growth of hair on the upper lip and then on the chin. Patches of pigmentation appear on the skin, particularly on the backs of the hands. There is a certain amount of obesity, or at least the fatty tissues are redistributed. The figure thickens, the waistline is lost, and the subcutaneous fat on the abdomen and thighs thickens. Muscle tone is reduced, particularly in the abdominal wall. Body weight changes during menopause may be related to decreased thyroid function, decreased physical activity, or increased food intake as a substitute for emotional satisfactions.

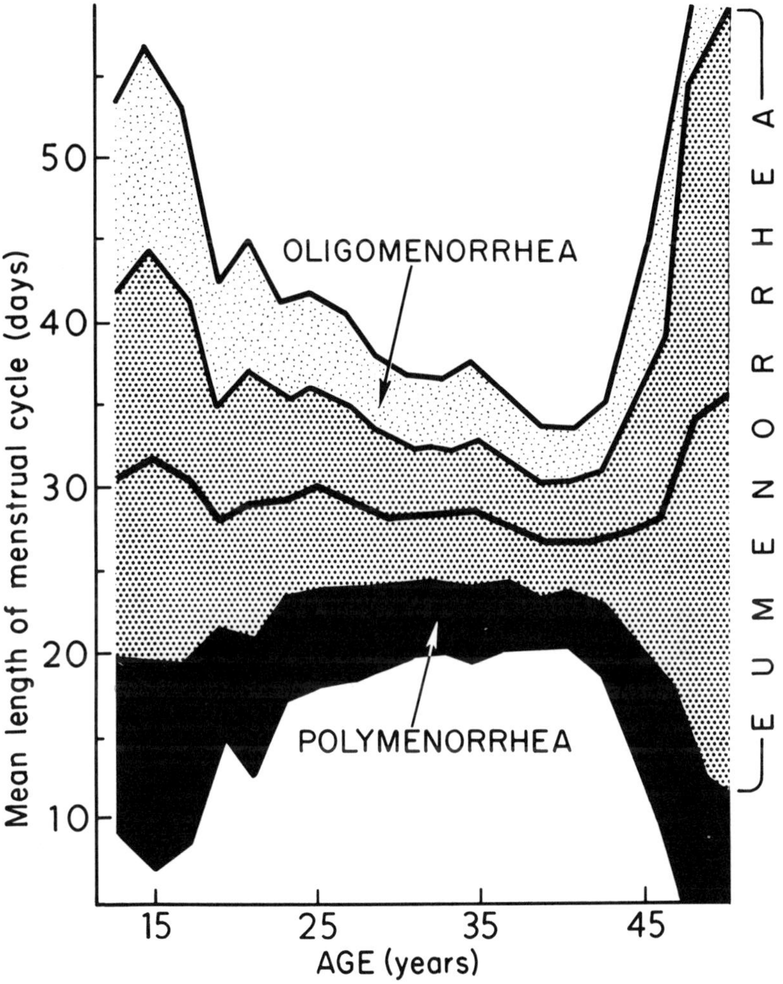

Figure 62. Menstrual performance correlated with age of women: M ± s = eumenorrhea; M + 2s = oligomenorrhea; M - 2s + polymenorrhea. M = average length of menstrual cycles; s = SD. (From Vollman, R. F. *Gynaecologia* (Basel) 142:310, 1956.)

Changes in Reproductive Organs

Depletion of ovarian oocytes begins before birth and proceeds at a rate partly controlled by hereditary factors. By the time of menopause, or soon after, few oocytes remain in the ovary. This depletion is closely associated with the termination of reproductive capacity. Cessation of ovulation occurs first, followed by cessation of ovarian hormonal functions. The decline in ovarian hormone production causes dramatic increases in gonadotropin secretion.

The postmenopausal period is characterized by several morphological and physiological changes (Figure 63; Table XXXI) which affect sexual behavior and reproductive phenomena:

1. The vaginal epithelium becomes thin and atrophic in response to estrogen deficiency. The labia minora and prepuce become atrophic, exposing the clitoris. The external genitalia become easily traumatized and frequent or prolonged intercourse becomes painful. Atrophic vaginitis seems to be common in postmenopausal women with an active coital life.
2. There is remarkable reduction in the amount and rapidity of onset of vaginal lubrication. A reduction in the soft tissue around the genitalia causes a loss of effective sexual stimulation due to lack of localized vasoconstriction.
3. The uterus becomes spastically contracted and lacks contraction associated with orgasm.

Vaginal changes associated with aging are hormone-dependent. Treatment with estrogen reverses the changes in vaginal walls and epithelium.

Menopause and Sexuality

The degree of sexual desire in menopausal women varies a great deal and is not related to the process of menopause. Frequently women acquire an increased desire for sexual gratification after the possibility of pregnancy has disappeared. Women 50 to 75 years of age retain the capacity for multiorgasmic response.

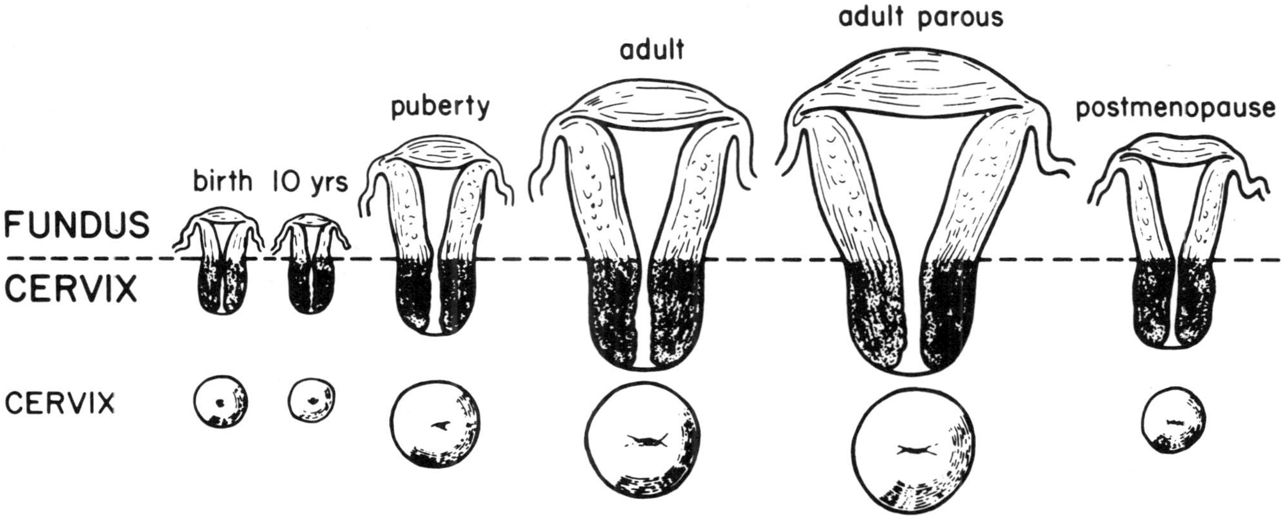

Figure 63. Changes in the relative size of uterus and cervix with advancing age. [Adapted and redrawn from Fluhmann, C. F. *The Cervix Uteri and Its Diseases* (Philadelphia: W. B. Saunders, 1961); Parsons and Sommers. *Gynecology* (Philadelphia: W. B. Saunders, 1969); Hafez, E. S. E. *Comparative Anatomy of the Mammalian Cervix*, R. J. Blandau and K. S. Moghissi, Eds. (Chicago: University of Chicago Press, 1973).]

Table XXXI

Physiological Changes in Reproductive Organs Associated with Aging

Organ	*Changes*
Vulva	Labial folds less prominent; skin becomes thinner; vascular and elastic fibers disappear; dermal vessels decrease
	Hair less abundant on mons pubis and labial folds; vulvo-vaginal glands decrease in number and secretory activity; subcutaneous fat is lost
	Atrophic changes extend to perineal body, rectum and anus; loss of tone of sphincter causes mild fecal incontinence
Vagina	Canal is shortened; thinning of vaginal epithelium leads to loss of vaginal rugae and reduced vascularity
	Vaginal wall is easily lacerated and irritated by bacterial or other noxious agents; lactobacilli are less frequent
	Posterior vaginal fornix flattens due to shrinkage
	Vaginal secretions become scanty, with higher pH
Uterus	Decreases in size due to fluid loss and transformation to interstitial fibrous tissues
Endometrium	Fewer endometrial glands; fewer ciliated cells with shorter kinocilia; reduction in size and number of microvilli of secretory cells
Myometrium	Less demarcation between endometrium and myometrium; myometrial fibers resemble fibrous tissue with hyperplasia and myoma
Cervix	Myometria blood vessels become thick and prominent; reduction in size is less than that of fundus so cervical-fundal relationship reverts to that of adolescence

	Squamocolumnar junction is pulled upward in cervical canal Cervical canal reduces in size; pyometra occurs in case of obstructed canal
Oviduct	Becomes shorter and thinner; mucosal folds become low, thin and constricted Endosalpinx thins, flattens and loses cilia Smooth muscle cells replaced by connective tissue
Pelvic Floor	Muscle tone declines Ligaments and connective tissue lose strength and elasticity Pelvic organs relax
Ovary	Irregular structure and function of corpus luteum; decreased progesterone; shortened luteal phase Ovulation fails; estrogenic deficiency is gradual, lasting 1 to several years
Mammary Glands	
Premenopausal	Subcutaneous fat increases; irregularities in lobules; cystic dilation; proliferation of lobular buds
Postmenopausal	Subcutaneous fat reabsorbed, parenchymatous tissue atrophies; lobules disappear; tubules collapse; fibrous stroma condense Nipples become smaller and less erectile; pigmentation decreases; breasts become flattened and pendulous
Testes	Decrease in size and firmness; basement membrane and tunica propria of seminiferous tubules thicken; tubules decrease in diameter; degenerative changes in germinal epithelium reduce sperm production Progressive intertubular fibrosis causes separation of seminiferous tubules from blood supply Sertoli cells contain golden-brown pigment granules
Prostate	Enlarges; its contractions become weaker

Table XXXII

Effect of Aging on Sexual Responses in Men and Women

Phase	*Climacteric Men*	*Postmenopausal Women*
Precoital Excitement	Slow penile erection; less scrotal vasoconstriction	Time required to lubricate vagina increases to 5 min (0.5 min for young women)
Plateau	Minimal vascular engorgement of testes; penile erection not associated with emission of pre-ejaculatory fluid	Lack of deep coloration of sex skin in minor labia; major labia do not elevate and flatten against perineum as in young women
Orgasm	Short duration; ejaculatory thrust comprises 1 or 2 contractions (4 to 6 contractions in young men)	Reduced duration of orgasm, reduced rhythmic uterine contractions may be spastic and pain-inducing
Refractory	Erect penis turns to flaccid state after ejaculation within a few seconds (minutes or hours in young men); refractory period is prolonged	Pelvic viscera returns to prestimulatory state very rapidly; expanded vagina collapses rapidly

Several factors may cause psychosexual dysfunction during aging (Table XXXII), *i.e.*, inadequate endrogenous sex steroids during menopause may cause painful intercourse due to atrophy of the vaginal epithelium mucosa and painful uterine contractions during orgasm. Also, vaginal lubrication becomes less and requires longer time during menopause; and reduced sexual interest of the partner is due to waning physical attractiveness as a result of weight gain and poor cosmetic care.

Decline in Fertility

Attempts to determine the time of cessation of reproductive ability are complicated by the widespread use of contraceptives and the influence of socioeconomic factors. The extreme age for successful pregnancy in women is about 52 years. Declining fertility in aged women (Figure 64) is partly due to less-frequent ovulation. In women aged 40 to 45 years, 70% of menstrual cycles are ovulatory and the percentage drops to 60% after 45 years of age. Postmenopausal pregnancies are rare. Conception may occur after 5 to 12 years of amenorrhea, but this is not necessarily related to menopause. It is estimated that 1 out of 20,000 to 60,000 births occurs past the age of 50. A Scottish woman had six successful pregnancies and one abortion after the age of 47, with the last child being born when the mother was 62.

Pregnancy in elderly multigravidas is generally hazardous to both the mother and child. There is an increased rate of abortion, stillbirth, obstetrical difficulty, congenital anomalies and maternal death.

Perinatal mortality increases steadily after 35 years of age and doubles over the age of 45. Some of the complications may be associated with high parity. Mothers of nine or more children have three times as many stillbirths as women with fewer previous pregnancies. High parity also increases maternal and perinatal morbidity and the risk of thromboembolic complications. Defects of the aged egg and inadequacies of the aged uterus are partly responsible for the increased reproductive wastage.

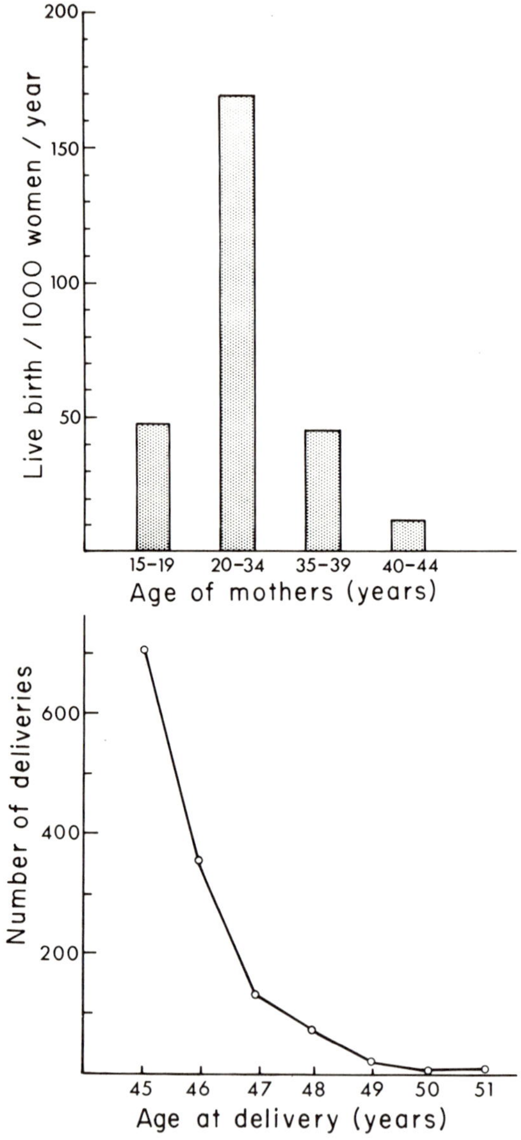

Figure 64. Effect of age on fertility in England and Wales: *Top.* Live births per 100 women per year. *Bottom.* Deliveries in women more than 44 years of age. (Data from Registrar General's Statistical Review of England and Wales for 1969. HMSO; Francis, J. A. *J. Reprod. Fertil. Suppl.* 12:89 1970.)

With an increase in maternal age, there is a high frequency of defective offspring with nondysjunctional chromosomes. The high frequency of abnormal fertilization in premenopausal women is due to the increasing opportunity for endogenous and exogenous factors to produce adverse chromosomal and perhaps cytoplasmic changes in the oocytes prior to ovulation. In women and nonhuman primates, ovulation is not associated with a clearly defined period of estrus (heat) as in animals. Thus, sperm may fertilize an egg which has been in the oviduct for a variable length of time, yet is still viable. Since premenopausal women experience less-frequent sexual intercourse, the changes of an egg becoming overripe before fertilization are increased, due to the less-frequent presence of viable sperm. There is also a possible relationship among frequency of intercourse, delayed fertilization and the incidence of mongolism (Down's Syndrome) due to trisomy of chromosome-21 in the offspring (Figure 65). Thus, aborted fetuses from older mothers show a higher percentage of abnormalities than those from younger mothers.

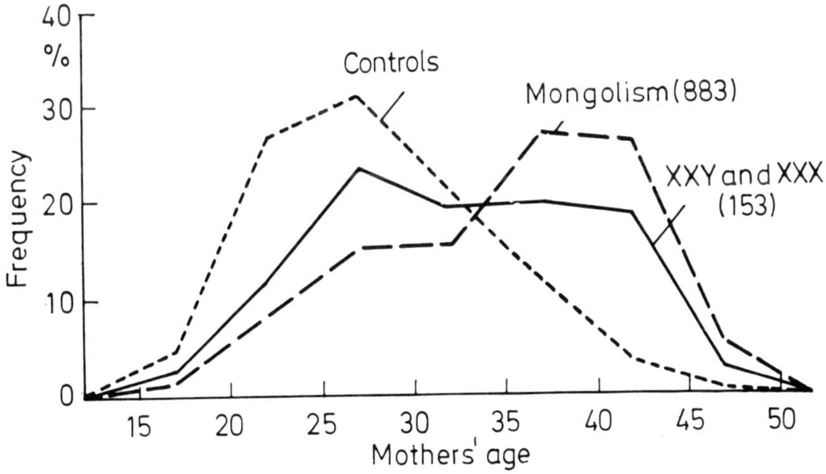

Figure 65. Age of mother on incidence of Down's Syndrome and XXY types (Klinefelter) and XXX (Triple X women) in comparison to normal population. [From Penrose, L. S. *Outline of Human Genetics*, 2nd ed. (London: Heinemann, 1963); Fuhrmann and Vogel. *Genetic Counseling* (New York: Springer-Verlag, 1969).]

Endocrine Mechanism

Menopause is associated with disturbances in the hypothalamo-pituitary-ovarian functions.

Estrogens

With aging, the ovary loses sensitivity to gonadotropins. The disappearance of growing follicles in the ovary causes sudden decline in estrogen secretion. Incomplete development or premature involution of corpora lutea results in short cycles and failure of pregnancy during early stages. Waves of estrogen secretion continue after ovulation has ceased, but become inadequate to stimulate the endometrium or to inhibit the release of pituitary gonadotropin. With atrophy of the endometrium, menstruation ceases and menopause occurs.

The small amounts of estrogens secreted by the adrenal gland may result in occasional bleeding. The ovarian stroma of menopausal and postmenopausal women secrete androgens, *e.g.*, androtenedione and testosterone. The endocrine effect of the postmenopausal ovary may be related to its contribution of androgens to the plasma pool. Postmenopausal women produce estrone derived from the aromatization of plasma androstenedione of andrenocortical origin. Some postmenopausal women who produce sufficient quantities of estrone from androstenedione do not experience estrogen deficiency after the menopause; such subjects do not need replacement therapy.

Menopausal estrogen therapy is beneficial for most women who experience estrogen deficiency, as judged from their vaginal cytology, within a year after cessation of menstruation. Indications for estrogen therapy include estrogen deficiency (vasomotor symptoms, atrophic genital changes, or hypogonadism) and osteoporosis whereas contraindications for estrogen use include vaginal bleeding, breast and endometrial carcinoma, liver disease, thrombosis and fibroids. Metabolic studies indicating negative calcium balance as a result of estrogen deficiency suggest that hormonal therapy is desirable.

Gonadotropins

The sudden disappearance of ovarian estrogen and the remarkable rise in plasma and urinary gonadotropin are used for clinical diagnosis of menopause. Both gonadotropic hormones increase equally, so that the premenopausal ratio of FSH to LH is maintained at about 2 to 1. The rise in gonadotropin level is due to increased secretion rate of the pituitary in response to a decreased negative feedback of estrogen from the ovary on the hypothalamus. Implants or injections of estrogens reduce gonadotropin levels.

MALE CLIMACTERIC

Men seem to age faster than women until 50 years of age. From age 30, black men age an average of 5 years more than white males; this difference increases after 50 years of age.

Sexuality of Aged Men

Sexual interest and capacity frequently extend into old age (Figure 66). The capacity of penile erection can be maintained until 75 years of age when 50% of men may engage in satisfactory sexual intercourse. About one-quarter of the male population becomes impotent by age 70. A large percentage of impotence is due to psychogenic factors. Some men, as they become old, believe or are made to believe that they are sexually impotent. This belief is reinforced by the inability to attain or maintain an erection during a few successive sexual attempts.

Advanced age is associated with dramatic changes in endocrine function and sexual behavior: slow arousal with general delay in sexual response, reduced erection, slow and decreased forceful ejaculatory thrust with fewer expulsive contractions, increased refractory periods between erections and between orgasms, decreased frequency of intercourse, prolongation of the sex flush, depression and paranoia (Table XXXII). Erection is delayed in aged men just as adequate vaginal lubrication takes

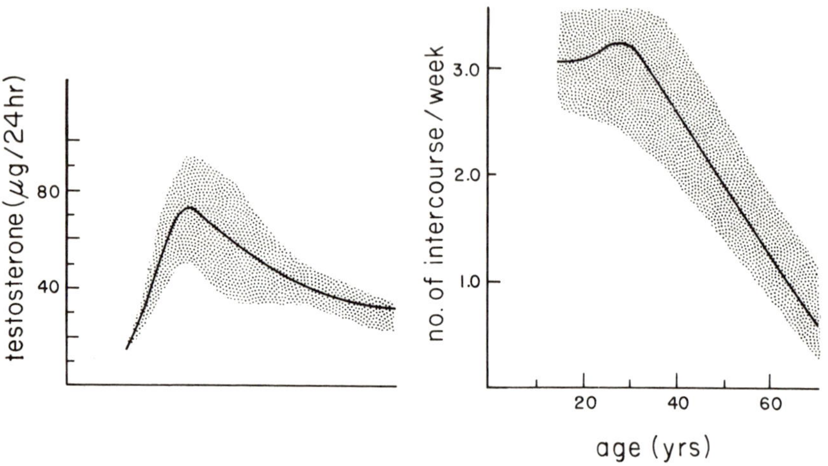

Figure 66. *Top.* Age changes in testosterone level in men. *Bottom.* Age changes in frequency of orgasm of single, married and previously married men. Shaded area indicates degree of variability in these phenomena. Note the peak of both curves at 30 years of age.

longer to achieve in postmenopausal women. These two phenomena are comparable, since penile erection and vaginal lubrication are regulated by the same nerve and vascular supply.

Different components of sexual behavior decline at different rates. The component that apparently declines most dramatically is the capacity for repeated orgasm within a short period of time, which reaches its peak in preadolescence and declines very rapidly after 20 years of age. The slowest declining component is frequency of extramarital intercourse; this behaivor may be influenced by increased "stimulus pressure."

The decreasing levels of urinary androgens during old age are associated with decreasing testosterone production and waning sexual activity. The extent to which waning sexual activity is determined by decreasing testosterone production is uncertain. Activity may decline because of decreased sensitivity in the target organs and other physiological impairment.

The psychological trauma of prostatectomy (a common operation in aged men) may lead to impotence whereas the surgical trauma of transurethral prostatectomy has no effect on penile erection but may cause retrograde ejaculation in the bladder.

Aging and Sperm Production

The aging process in the male gonad is less uniform and less remarkable than that in the female gonad. Aging is associated with several degenerative changes in the testes, which may appear as early as the third decade of life, and increase in severity with advancing age. Degenerative changes in the testis are also induced by disease and several environmental factors, the effects of which accumulate with time. Sperm production decreases and may cease entirely, although abundant sperm are often found in the testis of men in extreme old age. Although fertility is reduced, there are reliable reports of men becoming fathers at the ages of 70, 80 and even 90. Dramatic increase in the basement membrane of seminiferous tubules causes decreased spermatogenesis. Degenerative changes also occur in ejaculate volume, concentration of sperm, total number of sperm production, percentage of nonviable and abnormal sperm, physiological and biochemical characteristics of seminal plasma.

GENERAL EFFECTS OF AGING

The incidence of specific diseases associated with aging is partly genetic. Osteoporosis, which develops frequently after menopause, is associated partially with back pains. Its manifestations include loss of bone mass, increased bone porosity and decreased thickness of trabeculae and cortex. Most patients have relatively low calcium intake for several years. However, the principal defect seems to be a failure to replace protein lost during maintenance. This anabolic failure is due to poor protein intake; decreased levels of potent anabolic agents, estrogen and androgen; and the relative excess of catabolic adrenal corticosteroids which continue to be secreted at about the same rate as before menopause. The change may be due to restraint on osteoblastic activity accompanied by decreased vascularity rather than altered steroid levels. Trace elements probably play important roles in the pathogenesis of certain diseases that are common in the aged.

Extensive studies have been conducted to understand the biological mechanisms associated with aging. Several changes are not related to specific disease processes: tissue desiccation; retardation of cell division, capacity of cell growth and tissue repair, including capacity to produce immune bodies; lowered rate of tissue oxidation and decreased basal metabolic rate; cellular atrophy, degeneration, increased cell pigmentation and fatty infiltration; decreased speed, strength and endurance of skeletal muscle; degeneration and atrophy of nervous tissue; and impairment of homeostasis. Thinning and wrinkling of the skin due to loss of elasticity and atrophy of subcutaneous tissue are known effects of aging.

CHAPTER 13

FERTILITY REGULATION

MALE CONTRACEPTION

Male contraception has not been given much attention compared to the enormous volume of research devoted to female contraception. At present, male contraception is limited to voluntary bilateral vasectomy, the use of the condom and coitus interruptus. The ultimate aim in fertility regulation in males is the development of acceptable reversible methods that interfere with fertilizing capacity without compromising sexual libido.

Vasectomy

Vasectomy (surgical contraception), is a simple operation performed rapidly as an outpatient procedure. The vas deferens is readily identified by palpation before incising the skin. The success of the operation can be easily checked by semen analysis. There is a discrepancy concerning the number of ejaculates that should occur or length of time that should elapse before sterility is produced. Sperm seem to disappear from the ejaculate after 10 specimens have been collected, regardless of the time after vasectomy.

Antibodies to sperm are found in some healthy fertile men but more frequently in infertile men. The incidence and titers of such antibodies to sperm rise following vasectomy. Vasectomy does not seem to exert long-term harmful psychological or behavioral effects. A major drawback of vasectomy is its relative irreversibility. The microsurgical procedure for vasectomy reversal (vaso-vasostomy) requires careful reanastomosis of the vas deferens, and subsequent pregnancy rate is 20%. Methods are being developed for nonsurgical vasectomy.

Condoms

Condoms are made from skin or latex. Those made of latex are the most widely used and they come in different sizes, dry or lubricated, opaque, transparent, colored and in different shapes. In lubricated types, there is less incidence of rupture, due to reduced friction.

The condom is the ideal contraceptive for the couple with irregular intercourse and where unplanned coitus takes place. It is the only device that protects a couple against venereal diseases, and in such instances, it should be used during foreplay as well. It is used during treatment of Trichomonas vaginitis to prevent reinfection. It may be used in men with premature ejaculation as it may dull sensation and therefore prolong intercourse.

Coitus Interruptus

In coitus interruptus, the man, becoming aware of the imminence of ejaculation, withdraws his penis from the vagina before ejaculation. Seminal emission must occur away from the vagina and external genitalia of the female. Coitus interruptus requires considerable self-control, since there is an increased desire to achieve deeper penetration at the time of impending orgasm. Pre-ejaculatory lubricative fluids, which often ooze from the penis in response to sexual arousal, do not seem to contain sperm. However, this lubricating fluid may contain sperm after a recent previous ejaculation. Thus, urination is recommended between multiple acts of intercourse to flush out the sperm from the urogenital tract.

Several men are physiologically or psychologically unable to practice coitus interruptus. In certain men, the psychological pressures of coitus interruptus may result in impotence or premature ejaculation. The woman may fail to reach orgasm. Long-term practice of this method may lead to male impotence, tension in the couple and consequent deterioration of sexual relationships. Coitus interruptus is more popular in Catholic countries than in the Orient. In certain Eastern European countries where coitus interruptus is the major contraceptive method the lowest birth rates in the world have been achieved.

Experimental Approaches

Several experimental approaches have been used to control male fertility in laboratory animals, including nonhuman primates (Table XXXIII).

Disruption of Spermatogenesis

The spermatogenic process is precisely timed, and predictable periods of sterility can be induced by hormonal or pharmacological interference. The spermatogenic cycle lasts 10 to 11 weeks with an additional 2 weeks for sperm maturation and transport in the epididymis. Disruption of spermatogenic activity therefore results in the absence of sperm in the ejaculate within 4 to 12 weeks, depending on the particular spermatogenic cells affected by treatment.

There are two types of compounds that can cause disruption of spermatogenesis: (1) compounds that interfere with cell division during spermatogenesis, such as colchicine and heterocyclic agents, and (2) drugs that interfere with the hormonal milieu within the testes (Table XXXIV). For example, estrogens, androgens, progestogens, combinations of estrogen-androgen, or progestogen-androgen administered orally, intramuscularly, or by silastic subdermal implants, can interfere with the release of gonadotropins from the anterior pituitary. Antiestrogens, clomiphene citrate, and nonsteroidal gonadotropin inhibitors have also been effective.

Androgens and estrogens are normally produced in the testis. Testosterone is probably necessary for normal spermatogenesis whereas the role of estrogens is still uncertain. Androgens or estrogens induce testicular changes similar to those produced by hypophysectomy, *i.e.*, atrophy of the interstitial tissue and arrest of spermatogenesis at the primary spermatocyte or spermatogonium

Table XXXIII

Possible Methods and Mode of Action for Male Contraception

Target Organ	*Possible Contraceptive Methods*	*Mode of Action*
Hypothalamus-Pituitary	Administration of: androgen, antiandrogen, antigonadotropin, or antihypothalamic-releasing hormone	Inhibited biosynthesis or secretion of testosterone, gonadotropins and/or hypothalamic-releasing hormones Altered positive or negative feedback mechanisms
Testes-Epididymis	Administration of: androgen, estrogen, progesterone, cryproterone acetate, or pharmacological and immunosuppressant drugs	Altered spermatogenesis Delayed maturation of sperm in epididymis Delayed shedding of cytoplasmic droplet Inhibited development of directional motility necessary for fertilization Immobilization or cytotoxity of sperm upon ejaculation
Vas Deferens	Vasectomy (reversible or irreversible), occlusive devices (tantum clip)	Prevent sperm transport in male reproductive tract

Table XXXIV

Summary of Compounds Used to Disrupt Spermatogenesis by Interfering with Cell Division and/or Interfering with Hormonal Milieu in the Male

Mode of Action of Pharmacological Groups	*Compounds*
Gonadotropin Inhibitors	Steroids: estrogen, androgen with or without progestogens Nonsteroids: methallibure, chlorpromazine inhibitor, enheptin triethylamine hydrochloride Cyproterone, cyproterone acetate, α-chlorhydrine Amino-x-chlorhydrin (analogue) Inhibin (selective inhibition of FSH)
Antiandrogens	3-P-methylxyhenlyl-4-methyl-7-hydroxy-coumarin, methylcholanthrene, 2-acetyl-7-oxo-dodeca-hydroxyphenanthrene
Alkylating Agents	Nitrogen mustards Ethyleneimine derivatives Methanesulfonic acid, aziridines
Antimetabolites	Ethionine, antifolic compounds
Highly Specific Antiserum (immunological approach)	Androgen-binding protein secreted by Sertoli cell Autoimmune factors causing sperm agglutinating antibodies β-subunits of FSH
Heterocyclic Compounds	Nitrofuantoin, enheptin, 5-chloro-2-acetyl-thiophene
Antibiotic	Chloramphenicol BIS (dischloroacetyl) diamines, dinitropyroles

stages with remarkable reduction in the number of spermatogonia. Testosterone inhibits spermatogenesis via the pituitary with inhibition of FSH secretion. The suppressive effects of testosterone administration on the testis result from inhibition of the pituitary through feedback mechanisms. After withdrawal of testosterone treatment, spermatogenesis is restored and the sperm count begins to rise after the tenth week, reaching normal or even higher levels. The full rebound takes several weeks.

Antiandrogens interfere with sperm maturation after spermiation. Progestogens act as inhibitors of sperm production, and the exogenous testosterone compensates for the androgen loss. A single injection of the steroid, Depo-Provera, causes profound depression of sperm count and motility probably by a direct effect on the testes. Some of the side effects, however, include decrease of libido and weakness of erection.

Extensive investigations are in progress to isolate and characterize testicular inhibin, a naturally occurring testicular hormone which selectively inhibits FSH secretion.

Disruption of spermatogenesis has been accomplished by treatment of the testes with ultrasonic vibrations or ionizing radiation. Ultrasonic treatment interferes with cell membrane relationships, tubular membrane functions, and electrolyte transport. Germ cells show remarkable sensitivity to ionizing radiation. The various cellular components of seminiferous tubules show different radio-sensitivity. The most easily damaged cells are the Sertoli cells and spermatogonia, whereas spermatocytes and spermatids appear more resistant. Sloughing from the germinal epithelium occurs as a result of damage to Sertoli cells.

Interference with Sperm Transport

Removable plugs, valves, and occlusive and nonocclusive devices have been inserted into the vas deferens in an attempt to interfere with sperm transport. In one technique, an occlusive device implanted into a longitudinal incision along the principal axis of the vas deferens is held in position by sutures. In another technique, the vas is transected and the two ends are connected by a device containing a valve to either permit or prevent sperm transport. Such valve-like devices would allow the passage of sperm at the wish of the individual under the control of the andrologist. Nonocclusive devices are implants that release spermicidal reagents. These devices would not interrupt the flow of sperm and are less likely to impair spermatogenesis.

FEMALE CONTRACEPTION

Figure 67 shows the pregnancy rates of women who practice unprotected intercourse during reproductive years. Thus, female contraception is required for some 30 years. The common methods of female contraception include oral contraceptives, postcoital contraceptives, long-acting progesterone injectables, intrauterine devices, ovulation detection methods, voluntary sterilization, vaginal spermicides and mechanical barriers (Figure 68).

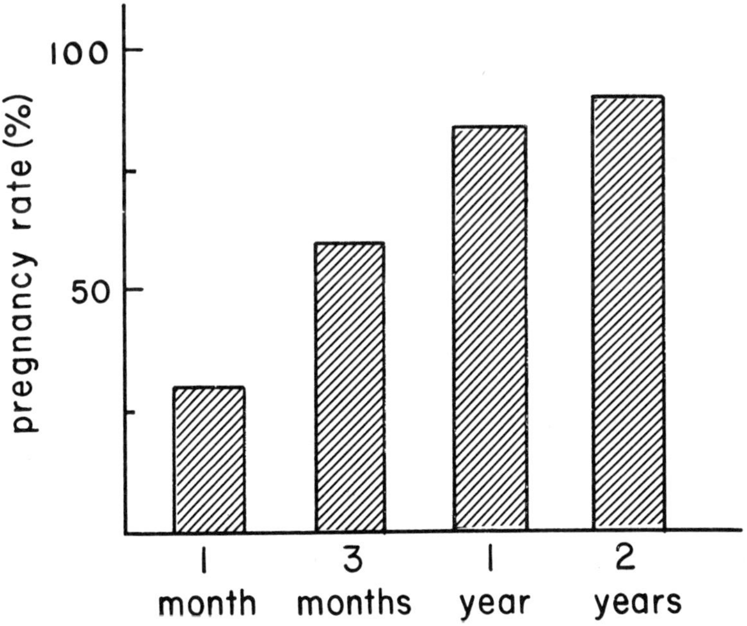

Figure 67. Pregnancy rate of women with unprotected coitus during reproductive years.

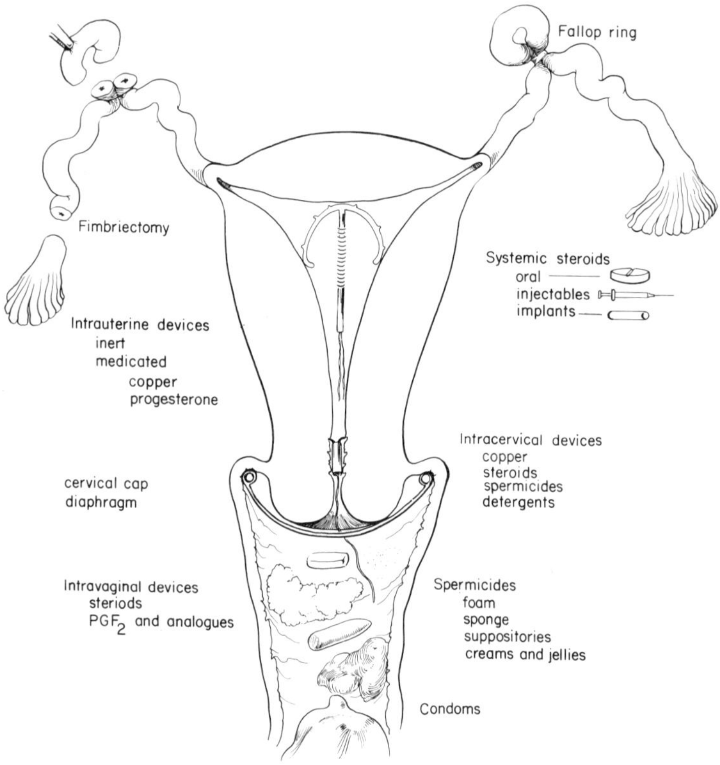

Figure 68. Various methods used for female contraception.

Oral Contraceptives

Wide application of steroid contraceptives has resulted from extensive research and modern advances in reproductive physiology, steroid chemistry, endocrinology, neuroendocrinology and radioimmunoassay for endocrine profiles. Several types of contraceptive steroids are available: combined estrogen-progestogen, sequential estrogen-progestogen, mini-pills containing progestogens alone or estrogen alone, and postcoital estrogens. The most commonly used pill in the USA is a combination of a synthetic estrogen and a progestogen. Combined contraceptives contain different doses of synthetic estrogens and progestins which vary in their potency, hormonal effects and side effects (Table XXXV).

Table XXXV

Common Oral Contraceptives Used in USA

Brand Name	*Estrogen*	*Dose (μg)*	*Progestogen*	*Dose (μg)*
Brevicon	Ethinyl Estradiol	35	Norethindrone	0.5
Demulen	Ethinyl Estradiol	50	Ethynodiol Diacetate	1
Enovid	Mestranol	75	Norethynodrel	5
Enovid-E	Mestranol	100	Norethynodrel	2.5
Loestrin	Ethinyl Estradiol	20	Norethindrone Acetate	1
Lo/Ovral	Ethinyl Estradiol	30	Norgestrel	0.3
Modicon				
Norinyl	Mestranol	50	Norethindrone	1
Norlestrin	Ethinyl Estradiol	50	Norethindrone Acetate	1
Norlestrin	Ethinyl Estradiol	50	Norethindrone Acetate	2.5
Nor-Q.D.			Norethindrone	0.35
Ortho-Novum	Mestranol Mestranol	80 100	Norethindrone Norethindrone	1 1
Ovral	Ethnyl	50	Norgestrel	0.5
Ovretts			Norgestrel	0.075
Ovulen	Mestranol	100	Ethynodiol Diacetate	1
Zorane	Ethinyl Estradiol	30	Norethindrone Acetate	1

These variations may be clinically important in certain patients. The two estrogens widely used are ethinyl estradiol or its 3-methyl ester, mestranol. The synthetic progestogens are related to either 19-nortestosterone or 17-hydroxyprogesterone. Combined contraceptives are usually taken as daily tablets for 21 consecutive days, starting when possible on the 5th day of the menstrual cycle. On completion of the 21-day course of tablets, medication is discontinued for 7 days.

Contraceptive steroids have also been used effectively for the relief of premenstrual tension and dysmenorrhea, menopausal symptoms, and the control of heavy or undiagnosed menstrual bleeding. There are various contraindications of the use of oral contraceptives: past history of thrombophlebitis, malignancy of the breast or reproductive system, pregnancy, history or presence of liver disease, prediabetic state or strong family history of diabetes, and mental retardation.

Mechanisms of Action

The action of oral contraceptives involves multiple physiological mechanisms. Many compounds, or even the same compound at various dosages, may have different modes of action. Compounds alone or in combination may exhibit multiple mechanisms of action. Oral contraceptives act on: (1) hypothalamus-pituitary-ovarian axis; (2) the endometrium; (3) cervical mucus; and (4) survival and transport of sperm and eggs (Figure 69). The relative effectiveness of these modes of action varies with the type of contraceptive.

Endocrine and Neuroendocrine Changes. Oral contraceptives prevent the release of pituitary gonadotropins required for ovulation. With the daily administration of 25 mg of norethindrone acetate, 50 μg of ethinyl estradiol, or 80 μg of mestranol, the midcycle LH peak is abolished. When both estrogen and progesterone are administered. they act on the hypothalamus and prevent the cyclic surge of LH. When mini-dosages of progestin are used, *e.g.*, 0.5 mg of chlormadinine acetate daily, the response is irregular; ovulation may occur, but conception is prevented by another mechanism.

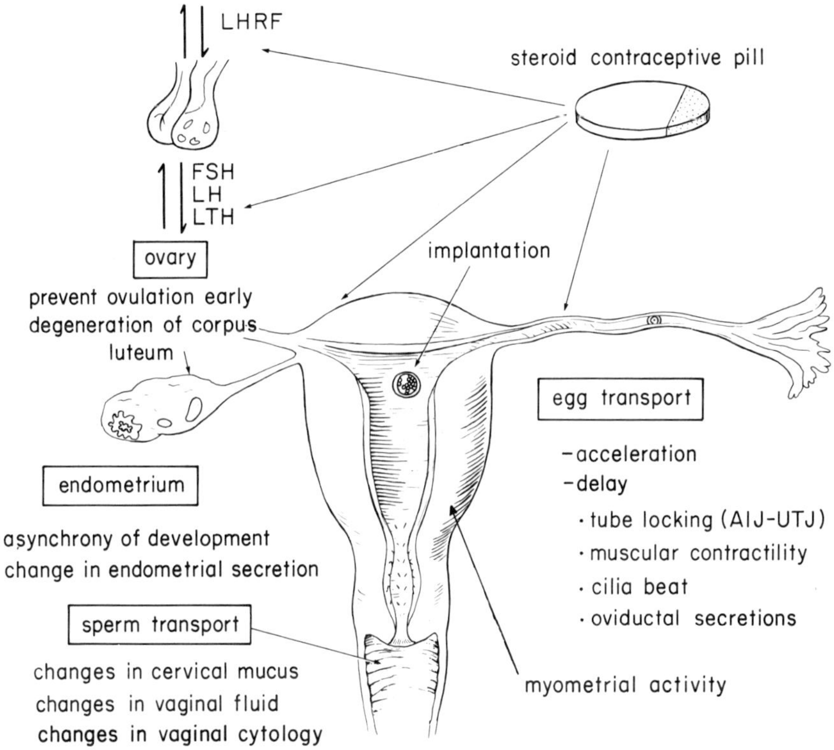

Figure 69. Diagrammatic illustration showing the mode of action of steroid contraceptives. Various physiological mechanisms may interfere with the sperm and egg transport, functions of the endometrium or myometrium, implantation of the blastocyst and/or hypothalamus-pituitary-ovarian axis.

Ovulation. Most combined oral contraceptives interfere with ovulation and ovarian steroidogenesis via suppression of the hypothalamo-hypophyseal system and inhibition of production and release of gonadotropins, especially the peaks of LH and FSH which normally occur at midcycle. The inhibition of ovulation causes structural and biochemical changes in the epithelium of the reproductive tract, physiochemical changes in oviductal fluid, endometrial fluid and cervical mucus, and altered motility of the oviduct and uterus. Prolonged administration of high doses of estrogen with or without progestogen reduces secretion of FSH and LH and causes morphological changes characteristic of depressed ovarian activity.

Cervical Mucus. Progestogens, alone or in combination with estrogen, inhibit mucorrhea (abundant watery cervical mucus); cervical mucus becomes scanty, cellular and highly viscous; ferning pattern disappears and the Spinnbarkeit is greatly reduced. The extent of changes in physical and chemical characteristics of cervical mucus depends on the type and dosage of hormone. Higher doses have a more remarkable and sustained effect than lower doses. Sequential contraceptives may exert no effect on cervical mucus, and may indeed improve its quality, making it more receptive to sperm penetration. Inhibition of sperm penetration through cervical mucus is an important factor contributing to the effectiveness of steroidal contraceptives and accounts in part for a lower pregnancy rate with combination than with sequential contraceptives.

Endometrium and Implantation of Blastocyst. Progestogens given throughout the menstrual cycle alter the physiological and biochemical characteristics of the endometrium and endometrial fluids, and luteal functions. The progestational changes take place in the endometrium at a faster rate causing it to be out of phase (asynchronous) with the ovulatory cycle. Predecidual changes in the uterine stroma and regression of the endometrial glands may prevent implantation of the blastocyst.

Mini-Pills. Mini-pills involve continuous administration of low doses of progestogen alone. Their mode of action is as follows: (1) the progestational cervical mucus interferes with sperm transport; (2) the disturbed ovarian steroidogenesis and an imbalance in the hypothalamus-pituitary-ovarian axis inhibit sperm and egg transport; and (3) the out-of-phase endometrium inhibits implantation of the blastocyst in the uterus.

Side Effects

The degree of pathophysiological, physical and psychological alterations in body functions as a result of steroidal contraceptives is still controversial. Throughout the normal menstrual cycle, cells of the vaginal epithelium undergo progressive maturational changes characterized by increased eosinophilia and flattening of cells. Maximal maturity is reached at the time of ovulation, whereas cellular maturity decreases after ovulation. The vaginal epithelium responds readily to local and systemic administration of steroid contraceptives. Estrogens exert remarkable effects on the vaginal mucosa. The basal layer of vaginal epithelium proliferates, matures and eventually becomes cornified and pyknotic. Different women show remarkable variation in the response of squamous epithelium to the same hormonal stimulation.

Contraceptive steroids have been associated with thrombosis, hypertension, and liver dysfunction. Changes in serum lipids may cause depression of lipoprotein lipase activity and stimulation of hyperinsulinism. Changes in liver enzyme levels are associated with a reduction in bile excretion. There are characteristic changes in plasma levels of amino acids, growth hormone, protein-bound thyroxine and cortisol; an increase in serum iron and copper; and a decrease in serum and urine magnesium. Depression is common in patients with a previous history of psychiatric symptoms.

These side effects have been described voluminously in the literature, but it is difficult to evaluate all the data. Unfortunately, there are no objective data of predictive value to enable physicians to distinguish those women who should or should not take oral contraceptives. Also, the long-term clinical significance of these side effects has not been established. Upon withdrawal of oral contraceptives, previous levels of fertility are restored except for rare cases of prolonged anovulation and postpill amenorrhea.

IUDs, spermicides, mechanical methods and sterilization have no known effect on lactation. Injectable contraceptives or oral progestogens have no adverse effect on lactation, and may augment milk production. However, oral contraceptives have an adverse effect on the quantity and possibly the quality of breast milk. Combined oral contraceptives should not be started until lactation is well established.

Postcoital Contraceptives

Postcoital contraception is the pharmacological inhibition of implantation after unprotected coitus. Theoretically, this is similar to "postconceptional" contraception, which is pharmacological interruption of pregnancy following implantation. A more convenient, yet practically unavailable method, is a "postovulatory" contraceptive agent administered following ovulation to prevent conception on a monthly regimen. Simple techniques for detection and timing of ovulation are mandatory for development of such contraceptives. A practical form would probably be pharmacological induction of menstruation (regardless of fertilization) by the administration of a compound in the second half of the menstrual cycle. The term "interceptives" was suggested to describe agents that interfere with implantation. Interceptives are neither contraceptives nor abortifacients since they are unable to inhibit fertilization or to interfere with an established implantation.

Estrogens, progestins and prostaglandins have been used as postcoital contraceptives in women. In an effort to prevent pregnancy following unprotected midcycle exposure, estrogen therapy is initiated within 24 hours and is continued for 5 consecutive days as follows: (1) diethylstilbestrol, 50 mg per day or ethinyl estradiol, 5 mg per day orally for 5 days; or (2) quingestanol acetate 0.8 to 1.5 mg or d-norgestrel 0.5 mg orally immediately following intercourse. These regimens are actually postovulatory rather than postcoital contraceptives.

The administration of estrogen at midcycle is ineffective should ovulation occur earlier or later in the cycle. Once implantation has occurred, the human pregnancy is remarkably tolerant to exogenous estrogens since administration of high doses of estrogen during the second week of gestation does not interfere with pregnancy.

Little is known about the effectiveness of prostaglandins in contraception. PGE_2 or $PGF_{2\alpha}$ are readily absorbed from the vagina and successfully induce menstruation within a few days after the first missed period. Prostaglandins may be administered vaginally once a month at the time of expected menstruation. However, prostaglandins may fail to prevent pregnancy due to: (1) an already established pregnancy; (2) prolonged time interval between unprotected intercourse and drug administration; (3) inadequate dose; (4) second coital exposure; (5) improper timing of drug administration relative to reproductive stage; (6) excessive vomiting; and (7) failure of the drug.

Postcoital antifertility effects may be mediated through interference with postovulatory events such as egg transport in the oviduct, fertilization, cleavage, viability of embryos, uterine physiology, implantation of the blastocyst and luteal function of the corpus luteum.

Modification of Egg Transport

Several pharmacological agents are effective in altering the patterns of egg transport in the oviduct. The ovum stays in the oviduct for three days during which time development of the fertilized egg is synchronized with that of the endometrium. Pharmacological interference with the physiologically developed synchrony inhibits implantation and opens new potentials for the control of fertility by interceptives.

Prostaglandins have a remarkable effect on oviductal motility. E-prostaglandins facilitate the passage of eggs to the uterus by virtue of their capacity to relax tubal occlusion, whereas F_2-prostaglandins delay egg transport by their occlusive potentials.

Myometrial and Endometrial Functions

On the transport of the egg from the oviduct to the uterus, the uterus undergoes certain developmental changes that are essential for survival and implantation of the embryo. Such changes include suppression of uterine contractility and progestational reaction of the endometrium. Provocation of myometrial activity at this stage may prevent implantation at a time before the menstrual cycle is disturbed and before pregnancy is established. The enzyme carbonic anhydrase seems to be implicated in implantation. Suppression of carbonic anhydrase content in the endometrium may also prevent implantation.

Luteolysis

Inhibition of progesterone synthesis by the corpus luteum is a potential means for preventing implantation. Constriction of blood vessels supplying the corpus luteum, and interference with hypothalamo-hypophyseal-ovarian axis cause luteolysis. Progesterone deficiency may follow direct or indirect luteolysis. In women exposed to unprotected intercourse around their expected time of ovulation, administration of a luteolytic substance a few days before their expected date of menstruation may result in normal menstrual flow whether or not pregnancy has occurred.

Effects on Early Embryo

The zona pellucida is shed shortly before attachment and implantation of the blastocyst. Inhibition of zona shedding prevents the embryo from coming in contact with the endometrium and interferes with implantation.

The enzyme monamine oxidase (MAO) is produced by the early embryo. It acts by breaking down 5-hydroxy-treptamine (5-HT) which is a metabolic by-product with toxic effects on the embryo. Inhibition of MAO synthesis causes accumulation of 5HT with subsequent termination of pregnancy. MAO inhibitors exert direct or indirect luteolytic effects.

Long-Acting Progesterone Injectables

Medroxyprogesterone acetate (DMPA) (Depo-Provera) is a long-acting progesterone injected (150 mg) intramuscularly every three months. The continuous release of progesterone inhibits the midcycle LH surge, thereby preventing ovulation. Depo-Provera alters the physical properties of cervical mucus, thus reducing sperm penetration. The endometrium undergoes remarkable atrophic changes due to the persistent antiestrogenic effects of progestogens. While a breakthrough ovulation may occur, the changes in cervical mucus and in the endometrium can either inhibit fertilization or prevent implantation. Upon withdrawal of injectable progestogens, ovulation is usually restored after several months.

Intrauterine Devices

Various shapes and sizes of intrauterine devices made of steel, copper or polyethylene are placed into the uterine cavity. A nylon "tail" string can be felt by the patient at the cervical os. The precise mechanism of action of IUDs in women is not fully known, although several theories have been proposed. IUDs produce a local inflammatory response. IUD-induced mobilization of polymorphonuclear leukocytes in the uterus may create a hostile environment to sperm and/or blastocyst and may be responsible for many of the biochemical and functional changes in the endometrium. IUDs also promote the accumulation of macrophages which engulf sperm through phagocytosis. The IUD induces subtle structural changes which may alter the biochemical characteristics of the endometrium. It is possible that IUDs produce lysozyme, which has blastolytic activity that interferes with implantation. In experimental animals IUDs alter myometrial function, impair endometrial sensitivity to mechanical stimuli at the time of implantation, and inhibit fertilization and gamete transport.

Natural Family Planning (NFP)

There are remarkable differences in reproductive ability (as judged by fertilization) throughout the menstrual cycle. As ovulation is approached, there is a gradual increase in fertility reaching a maximum of 30% on the day before ovulation. Immediately after ovulation there is a sudden decrease in fertility to nil on the third day after ovulation, until subsequent menstruation. Natural family planning (NFP) methods and their efficacy, depend on reliable ovulation detection and prediction, and abstinence during the fertile phase of the menstrual cycle.

Clinical Methods of Natural Family Planning

Three classical methods have been used clinically for natural family planning: basal body temperature (BBT), Calendar-Rhythm method, and symptothermal method. The Calendar-Rhythm method is based on the tendency of ovulation to occur 12 to 16 days before the next menstrual period. With careful recording of menstrual cycles, a woman can predict her fertile period. The Calendar Rhythm method is the least reliable of all NFP methods. Although claims have related anocephaly, mongolism and fetal anomalies to NFP, there is only evidence to support that Calendar Rhythm may be associated with spontaneous abortion.

The BBT method (temperature chart) is more accurate than the calendar method for predicting the expected date of ovulation. There is a small rise in the basal body temperature at the time of ovulation (see Chapter 7). If this rise in BBT lasts for three days, it is safe to say that ovulation has taken place at the beginning of the rise in temperature, then fertilization is highly unlikely to occur during intercourse in the rest of the menstrual cycle. The BBT method (Figure 70) is based on the thermogenic properties of progesterone during the luteal phase. A biphasic BBT is indicative of ovulation, and the rise in temperature coincides with ovulation time. However, in a certain percentage of cycles, ovulation may not be associated with biphasic BBT as a result of technical errors, improper recording, or inability of an adequate level of circulating progesterone to exert its thermogenic effect (deficient luteal phase). Also, a monophasic BBT during an apparently normal menstrual cycle is not conclusive evidence of anovulation.

The symptothermal method monitors the temperature shift in conjunction with other objective and subjective symptoms of ovulation such as midcycle pain (Mittelschmerz), which most women can recognize at ovulation, midcycle vaginal blood seepage in the cervix, and Spinnbarkeit–a cascade of thread-like, clear, colorless cervical mucus at the time of ovulation. The cervical mucus appears most commonly from 19 to 14 days before the end of the cycle, the mode being 16, but there is an appreciable number of cycles outside this range.

Cervical Mucus for Natural Family Planning

Cyclic changes in quantity and characteristics of cervical mucus reflect ovarian function and are employed in natural family planning (Table XXXVI). Recognition of the "peak" day of

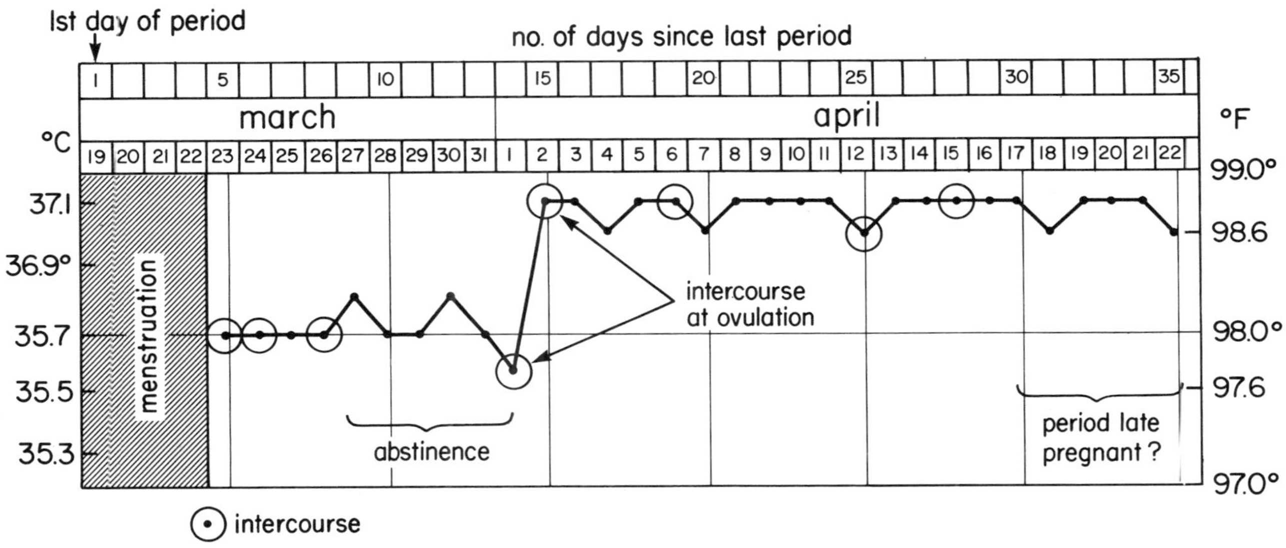

Figure 70. Natural family planning method. Note infertile and fertile periods. Fertilization of the egg seems possible during the fertile period when the association of sperm and eggs is possible around the time of ovulation.

Table XXXVI

Phases of Cervical Mucus as Used for Detection of Ovulation and Natural Family Planning

Phases of Menstrual Cycle	*Predominant Hormonal Level*	*Characteristics of Cervical Mucus*
Immediately before menstruation	Low estrogen	Minimal volume
Preovulatory	Rising estrogen	Scanty and sticky in consistency
Ovulation (midcycle and 2-3 days after ovulation)	Maximal estrogen	Maximal volume, thin and clear
Second half of cycle until next menstruation	High progesterone	Decreased volume, thick and sticky

"fertile mucus" is crucial to the determination of the postovulatory "safe" days. Extensive investigations have been conducted to correlate peak days of fertile mucus with endocrine profile, *e.g.*, estradiol, progesterone, LH and FSH.

"Cervical scores" is based on the amount of mucus ferning and Spinnbarkeit and on the condition of the cervix. The day of maximum cervical score is a reliable method of detecting ovulation, but requires daily examination by a paramedical worker; thus, it is of limited value for family planning. However, it is possible to teach patients to predict ovulation time by observing changes in their cervical mucus, *e.g.*, sensations of dryness and wetness, amount, color and consistency of mucus collected from the vulva after urination.

Several instruments have been developed to measure viscoelastic properties of cervical mucus for routine examination of patients. "Ovutime Eccentric Viscometer" measures rheological properties of biological and other non-Newtonian fluids, utilizing the "shear" of fluid, and the "resistance to flow." These two functions are combined into a single "yield point," when elastic flow is established by overcoming forces resisting flow. These methods are still in the experimental stage.

Laboratory Methods

Several laboratory methods have been used for the detection of ovulation, *e.g.*, the midcycle LH surge, progesterone rise in relation to basal body temperature and the endometrium, ascorbic acid excretion and vaginal smear. The time interval between the LH and FSH peaks and ovulation is 16 hours, and between the estradiol peak and ovulation is 36 hours. However, there is a wide range of time intervals among different women.

Biochemical changes in concentration of specific metabolites in vaginal secretions and cervical mucus as serially monitored have been used to predict and detect the time of ovulation. Electrophoresis was used to monitor alkaline phosphatase in cervical mucus, whereas gas chromatography was used to identify midcycle peaks of urea and lactic acid, which occur in the vaginal secretions 48 hours before to 24 hours after LH surge. However, the long-term prediction of ovulation by at least six days seems to be difficult and, as yet, unsolved.

Devices and Kits for Ovulation Prediction

Several devices have been developed to predict ovulation. An "electronic speculum" is designed to measure the changes in heat energy in the vagina as estrogen is produced, to detect the day of ovulation. A "pocket calculator" is used to replace the BBT thermometer, and to graph results, by a simple and convenient means. Such a simple instrument is capable of storing a sequence of previous readings and automatically detecting shifts and peaks. Another device, the "Ovumeter," predicts the time of ovulation by measuring changes in cervical mucus.

Various biochemical tests have been developed to measure shifts in the concentrations of various substances associated with ovulation in the urine, saliva, or cervical mucus. Most of these tests depended on the use of specially treated paper or tampons and proved impracticable. Some were too sensitive, others were not accurate. Today, none of these home testing kits are in use, although research continues in several countries.

Lactation and Postpartum Natural Family Planning

The initiation and maintenance of lactation are associated with a rapid decline of steroid levels after delivery and an increase in prolactin levels. Ovulation is inhibited during much of lactation. The regular stimulation of the nipple during breast feeding causes the release of chatecholamines in the hypothalamus, which, in turn, increases prolactin secretion by the anterior pituitary. Poor nutritional conditions of mothers associated with protein-deficient diets reduce milk volume. In such cases, the infant will have to suck harder to get his nutritional requirements. Excessive stimulation of the nipple will increase the anovulatory interval following delivery.

Menstruation returns in 1.3% of women by the end of the sixth week postpartum and its incidence increases gradually to 60% by the end of the first year. There is a gradual increase in the amount of menstrual blood and duration of flow during the first four cycles. The incidence of ovulation in menstruating lactating mothers is higher than in amenorrheic lactating mothers. Postpartum ovulation may occur 17 to 42 days after delivery according to extent and length of nursing.

Lactational amenorrhea may be used as the natural infertile period. The first postpartum ovulation must be diagnosed, and the corresponding sexual abstinence maintained. The cervical mucus method should be used for the ovulation detection during this period, since the BBT in the presence of nursing presents a biphasic pattern in less than 50% of the ovulatory cycles. The postpartum anovulatory period should not be considered as an absolute infertile period after 17 days postpartum. The risk of pregnancy increases with the length of lactation, and thus, some form of added contraception should be adopted during lactation. Contraceptive measures should be initiated one month postpartum in nonlactating women and four months postpartum in full-lactating women.

Table XXXVII

Methods of Female Sterilization

Surgical Approach		*Technique*	
Surgical approach to oviduct	Abdominal	Laparotomy	Incision to abdomen to deliver oviducts by hands
		Minilaparotomy	Raising uterus and oviduct into abdomen by a uterine elevator inserted vaginally, which allows a small incision
		Laparoscopy	Use of special endoscopic equipment inserted through abdominal wall
	Vaginal	Colpotomy	Operating under direct observation
		Culdoscopy	Use of an endoscope
	Cervical	Hysteroscopy	Occlusion of oviducts via cervix and uterine lumen
Sterilization approach	Direct observation	Oviduct tied or cut and tied with absorbable catgut	
		Chemical occlusion	
	Laparoscopy or direct observation	Electrocoagulation	
		Application of clips or Fallop rings	
	Ovariplexy	Burial of ovaries in broad ligament	

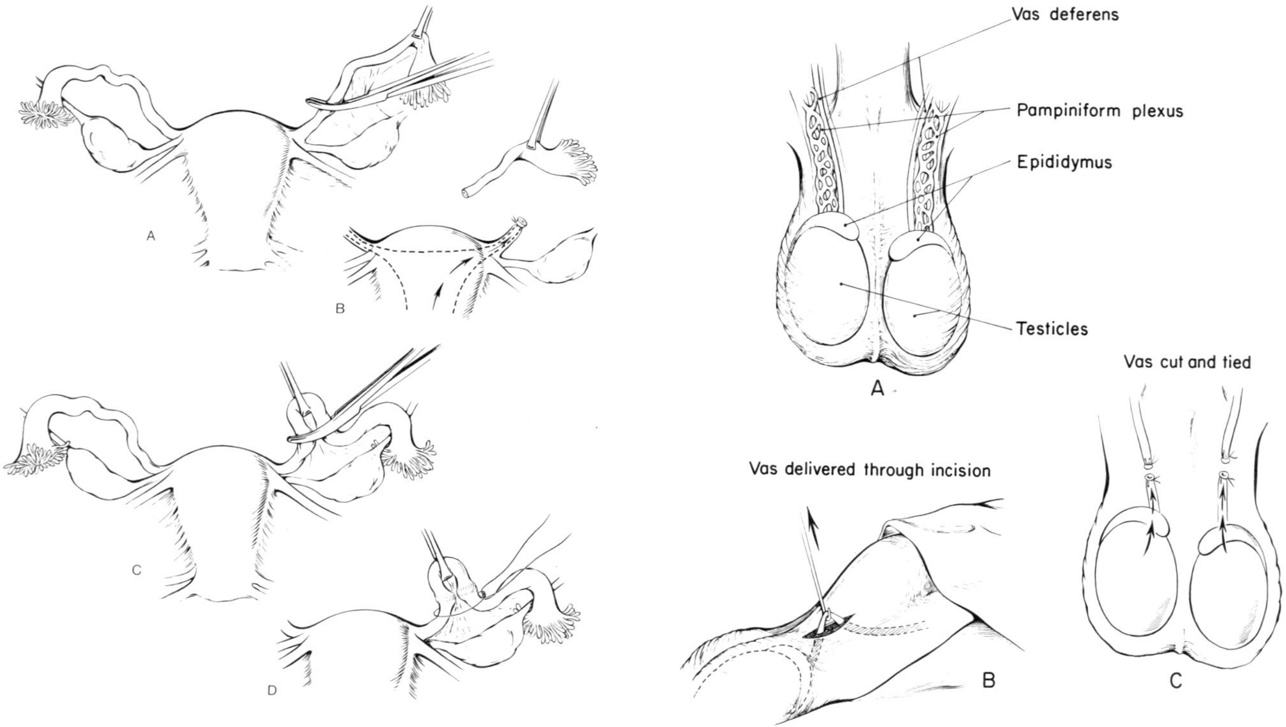

Figure 71. Surgical approaches to male (right) and female (left) sterilization (Hofmeister, 1976).

Female Sterilization

Several operative procedures are used by cutting or ligating the oviducts via the vaginal or abdominal wall using laparoscopy or laparotomy (Table XXXVII; Figure 71), or by elective vaginal or abdominal hysterectomy. In nonpregnant and postabortum women the oviducts lie in the pelvic cavity, whereas in post-partum women, the oviducts are drawn out of the pelvis and are easily accessible. The egg and sperm degenerate and are readily absorbed in the oviduct whether or not it is ligated.

Vaginal Spermicides

Several types of suppositories, jellies, creams, tablets and foams contain a variety of spermicidal compounds: *p*-tri-isopropylphenoxypolythoxyethanol, quinine, boric acid, lactic acid, ricinoleic acid, chinosol, hexylresorcinol, or formaldehyde. Vaginal spermicides do not harm the vaginal wall, but they act as a mechanical barrier against sperm penetration and lower the pH of the vagina, thus immobilizing and destroying sperm.

Female Barrier Contraceptives

Cervical caps and diaphragms ranging in size from 5 to 10 cm and coated with spermicidal agents adhere to the cervix to prevent sperm transport to the uterine lumen.

BIBLIOGRAPHY

BOOKS IN REPRODUCTIVE MEDICINE

Austin, C. R. and R. V. Short. *Reproduction in Mammals* (Cambridge, Mass.: Cambridge University Press, 1972).

Beller, F. K., K. Knorr, C. Lauritzen and R. M. Wynn. *Gynecology: A Textbook for Students* (New York: Springer-Verlag, 1974).

Diczfalusy, E. and U. Borell, Eds. *Control of Human Fertility* (Baltimore: Williams & Wilkins, 1971).

Hafez, E. S. E., Ed. *Reproduction in Farm Animals* (Philadelphia: Lea & Febiger, 1974).

Hafez, E. S. E., Ed. *Reproduction and Breeding Techniques of Laboratory Mammals* (Philadelphia: Lea & Febiger, 1970), p. 375.

Hafez, E. S. E., Ed. *The Mammalian Fetus: Methodology and Comparative Biology* (Springfield, Illinois: Charles C. Thomas, 1975).

Hafez, E. S. E., Ed. *Scanning Electron Microscopy Atlas of Mammalian Reproduction* (New York: Springer-Verlag, 1975).

Hafez, E. S. E., Ed. *Aging and Reproductive Physiology* (Ann Arbor, Michigan: Ann Arbor Science Publishers, Inc., 1976).

Hafez, E. S. E., Ed. *Human Ovulation: Regulation, Prediction, Detection and Induction* (Amsterdam: Elsevier/North Holland, 1978).

Hafez, E. S. E. and T. N. Evans, Eds. *Human Vagina* (New York: Elsevier/North Holland, 1978).

Hafez, E. S. E. and J. Reel, Eds. *Hypothalamic Hormones* (Ann Arbor, Michigan: Ann Arbor Science Publishers, Inc., 1975).

Hafez, E. S. E. and C. G. Thibault, Eds. *The Biology of Spermatozoa: Transport, Survival and Fertilizing Ability* (Basel, Switzerland: S. Karger, 1974).

Hamilton, D. W. and R. O. Greep, Eds. *Male Reproductive Organs, Handbook of Physiology*, Vol. V (Washington, D.C.: American Physiological Society, 1975), Section 7.

Hubinont, P. O. and M. L'Hermite, Eds. *Sperm Action* (Basel, Switzerland: S. Karger, 1975).

James, V. H. T., M. Serio and L. Martini, Eds. *The Endocrine Function of the Human Testes* (New York: Academic Press, 1973).

Moghissi, K. S. and E. S. E. Hafez, Eds. *Biology of Mammalian Fertilization and Implantation* (Springfield, Illinois: Charles C. Thomas, 1972).

Niswander, K. R. *Obstetrics: Essentials of Clinical Practices* (Boston, Mass: Little, Brown & Co., 1976).

Parkes, A. S. *Patterns of Sexuality and Reproduction* (New York: Oxford University Press, 1976).

Peel, J. and M. Potts. *Textbook of Contraceptive Practices* (Cambridge, Mass.: Cambridge University Press, 1969).

Sciarra, J. J., C. Markland and J. J. Speidel, Eds. *Control of Male Fertility* (New York: Harper & Row, 1975).

Weinberg, M. S. and C. J. Williams. *Male Homosexuals: Their Problems and Adaptations* (New York: Oxford University Press, 1974).

World Health Organization. "Methods of Fertility Regulation: Advances in Research and Clinical Experience," *WHO Tech. Rep. Serv.*, Geneva (1971).

MALE REPRODUCTIVE ORGANS

de la Balze, F. A., G. E. Bur, F. Scarpa-Smith and J. Irazu. "Elastic Fibers in the Tunica Propria of Normal and Pathologic Human Testes," *J. Clin. Endocrinol. Metab.* 14:626 (1954).

Heller, C. G. "Kinetics of the Germinal Epithelium in Man," *Rec. Progr. Horm. Res.* 20:545-575 (1964).

Lipsett, M. B., H. Wilson, M. A. Kirschner, G. A. Sarfaty and C. W. Bardin. "Studies on Leydig Cell Physiology and Pathology: Secretion and Metabolism of Testosterone," *Rec. Progr. Horm. Res.* 22:245-271 (1966).

Prader, A. "Testicular Size: Assessment and Clinical Importance," *Triangle* 7:240 (1966).

Zachmann, M. A., A. Prader, H. P. Kind, H. Hafliger and H. Budliger. "Testicular Volume During Adolescence: Cross-Sectional and Longitudinal Studies," *Helv. Paediat. Acta.* 29:61 (1974).

FEMALE REPRODUCTIVE ORGANS

Blandau, R. J. and K. Moghissi, Eds. *The Biology of the Cervix* (Chicago: The University of Chicago Press, 1973).

Hafez, E. S. E. "Endocrine Control of the Structure and Function of the Mammalian Oviduct," in *Handbook of Physiology*, American Physiological Society, *Endocrinology II*, Part 2, Chapter 35 (1973), pp. 97-122.

Hafez, E. S. E., H. Ludwig and H. Metzger. "Human Endometrial Fluid Kinetics as Observed by Scanning Electron Microscopy," *Am. J. Obstet. Gynecol.* 122:929-938 (1975).

Iacobelli, S., *et al.* "Biochemistry of Cervical Mucus: A Comparative Analysis of the Secretion from Preovulatory, Postovulatory and Pregnancy Periods," *Fertil. Steril.* 22:727-734 (1971).

Moghissi, K. S. "The Function of the Cervix in Fertility," *Fertil. Steril.* 23:295-306 (1972).

HORMONES OF REPRODUCTION

Hafez, E. S. E. and J. R. Reel, Eds. *Hypothalamic Hormones* (Ann Arbor, Michigan: Ann Arbor Science Publishers, Inc., 1975).

O'Malley, B. W. and A. R. Means. "Female Steroid Hormones and Target Cell Nuclei," *Science* 183:610.

Schally, A. V., A. J. Kastin and A. Arimura. "The Hypothalamus and Reproduction," *Am. J. Obstet. Gynecol.* 122:857 (1975).

SEXUAL MATURITY

Boyar, R. M., J. W. Finkelstein, R. David, H. Roffwarge, S. Kapen, E. D. Weitzman and L. Hellman. "Twenty-four Hour Patterns of Plasma Luteinizing Hormone and Follicle-Stimulating Hormone in Sexual Precocity," *New Eng. J. Med.* 289:282 (1973).

Hopper, B. R. and S. S. C. Yen. "Circulating Concentrations of Dehydroepiandrosterone and Dehydroepiandrosterone Sulfate During Puberty," *J. Clin. Endocrinol. Metab.* 40:458 (1975).

Illig, R. and A. Prader. "Effect of Testosterone on Growth Hormone Secretion in Patients with Anorchia and Delayed Puberty," *J. Clin. Endocrinol.* 30:615 (1970).

Job, J. C., P. E. Garnier, J. L. Chaussain, R. Scholler, J. E. Toublanc and P. Canlorbe. "Effect of Synthetic Luteinizing Hormone-Releasing Hormone (LHRH) on the Release of Gonadotropins in Hypophysogonadal Disorders of Children and Adolescents. V. *Agonadism*," *J. Clin. Endocrinol. Metab.* 38:1109 (1974).

Johanson, A. and E. Vann. "Fluctuations of Gonadotropin Levels in Children," *J. Clin. Endocrinol. Metab.* 39:154 (1974).

Judd, H. L., D. C. Parker, J. S. Rakoff, B. R. Hopper and S. S. C. Yen. "Elucidation of Mechanism(s) of the Nocturnal Rise of Testosterone in Men," *J. Clin. Endocrinol. Metab.* 38:134 (1974).

Lamberg, B. A., R. L. Kantero, P. Saarinen and O. Wedholm. "Endocrine Changes Before and After the Menarche. IV. Serum Thyrotropin in Female Adolescents," *Acta. Endocrinol.* 74:695 (1973).

Lee, P. A. and C. J. Migeon. "Puberty in Boys: Correlation of Plasma Levels of Gonadotropins (LH, FSH), Androgens (Testosterone, Androstenedione, Dehydroepiandrosterone and Its Sulfate), Estrogens (Estrone and Estradiol) and Progestins (Progesterone and 17-hydroxyprogesterone)," *J. Clin. Endocrinol. Metab.* 41:556 (1975).

Root, A. W. "Endocrinology of Puberty. II. Normal Sexual Maturation," *J. Pediatrics* 83:1 (1973).

Root, A. W. and E. Reiter. "Evaluation and Management of the Child with Delayed Pubertal Development," *Fertil. Steril.* 27:745 (1976).

Shearman, R. P. and J. R. Turtle. "Secondary Amenorrhea with Inappropriate Lactation," *Am. J. Obstet. Gynecol.* 106:818 (1970).

Speroff, L., R. H. Glass and N. G. Kase. "Puberty," in *Clinical Gynecologic Endocrinology and Infertility* (Baltimore: Williams & Wilkins, Co., 1973), p. 28.

Winter, J. S. D. and C. Faiman. "The Development of Cyclic Pituitary-Gonadal Function in Adolescent Females," *J. Clin. Endocrinol. Metab.* 37:714 (1973).

FUNDAMENTALS OF ANDROLOGY

Craft, I. and J. McQueen. "Effect of Irrigation of the Vas on Postvasectomy Semen Counts," *Lancet* 1:515 (1972).

Pardanani, D. S., N. G. Patil and H. N. Pawar. "Some Gross Observations of the Epididymides Following Vasectomy: A Clinical Study," *Fertil. Steril.* 27:267 (1976).

Peterson, N. T., A. R. Midgley and R. B. Jaffe. "Regulation of Human Gonadotropins. III. Luteinizing Hormone and Follicle-Stimulating Hormone in Sera from Adult Males," *J. Clin. Endocrinol.* 28:1473-1478 (1968).

Rowley, M. J. and C. G. Heller. "The Testosterone Rebound Phenomenon in the Treatment of Male Infertility," *Fertil. Steril.* 23:498 (1972).

MENSTRUAL CYCLE

Henzl, M. R. and E. J. Segre. "Physiology Human Menstrual Cycle and Early Pregnancy. A Review of Recent Investigations," *Contraception* 1:315-338 (1970).

Johansson, E. D., U. Larsson-Cohn and G. Gemzell. "Monophasic Basal Body Temperature in Ovulatory Menstrual Cycles," *Am. J. Obstet. Gynecol.* 113(8):933-937 (1972).

Knobil, E. "On the Regulation of the Primate Corpus Luteum," *Biol. Reprod.* 8:246 (1973).

Moghissi, K. S., F. N. Syner and T. N. Evans. "A Complete Picture of the Menstrual Cycle," *Am. J. Obstet. Gynecol.* 14:405-418 (1972).

Treolar, A. E., R. E. Boynton, B. G. Behn and B. W. Brown. "Variation of the Human Menstrual Cycle through Reproductive Life," *Internat. J. Fertil.* 12:77-126 (1967).

Vande Wiele, R. L., J. Bogumil, I. Dyrenfurth, M. Ferin, R. Jewelwicz, M. Warren, T. Rizkallah and G. Mikhail. "Mechanisms Regulating the Menstrual Cycle in Women," *Rec. Progr. Horm. Res.* 26:63 (1970).

Yussman, M. A. and M. L. Taymor. "Serum Levels of Follicle-Stimulating Hormone and Luteinizing Hormone and of Plasma Progesterone Related to Ovulation by Corpus Luteum Biopsy," *J. Clin. Endocrinol.* 30:396 (1970).

SURVIVAL AND TRANSPORT OF SPERM AND EGGS

Blandau, R. J. "Gamete Transport–Comparative Aspects; the Mammalian Oviduct," in *Comparative Biology and Methodology*, E. S. E. Hafez and R. J. Blandau, Eds. (Chicago: University of Chicago Press, 1969).

Hafez, E. S. E. "Gamete Transport," in *Human Reproduction: Conception and Contraception*, E. S. E. Hafez and T. N. Evans, Eds. (New York: Harper & Row, 1973).

Hafez, E. S. E. "Transport and Survival of Spermatozoa in Human Female Reproductive Tract," in *Human Semen and Fertility Regulation in Men*, E. S. E. Hafez, Ed. (St. Louis, Missouri: C. F. Mosby, 1976).

Soupart, P. "Sperm Capacitation: Methodology, Hormonal Control and Search for a Mechanism," in *Biology of Fertilization and Implantation*, K. S. Moghissi and E. S. E. Hafez, Eds. (Springfield, Illinois: Charles C. Thomas, 1972).

Thibault, C. "Physiology and Physiopathology of the Fallopian Tube," *Internat. J. Fertil.* 17:1 (1972).

FERTILIZATION AND IMPLANTATION

Boving, B. G. "Anatomy of Reproduction," in *Obstetrics*, J. P. Greenhill, Ed. (Philadelphia: W. B. Saunders, 1965).

Brackett, B. G., H. M. Seitz, G. Rocha and L. Mastroianni. "The Mammalian Fertilization Process," in *Biology of Mammalian Fertilization and Implantation*, K. S. Moghissi and E. S. E. Hafez, Eds. (Springfield, Illinois: Charles C. Thomas, 1972).

Edwards, R. G., P. C. Steptoe and J. M. Purdy. "Fertilization and Cleavage *In Vitro* of Preovulatory Human Oocytes," *Nature (London)* 227:1307 (1970).

Fridhandler, L. "Gametogenesis to Implantation," in *Biology of Gestation*, Vol. I, *The Maternal Organism*, N. S. Assali, Ed. (New York: Academic Press, 1968).

Hertig, A. T., J. Rock, E. C. Adams and M. F. Menkin. "Thirty-four Fertilized Human Ova, Good, Bad, and Indifferent, Recovered from 210 Women of Known Fertility. A Study of Biologic Wastage in Early Human Pregnancy," *Pediatrics* 28:202 (1959).

Seitz, H. M., G. Rocha, B. G. Brackett and L. Mastoianni. "Cleavage of Human Ova *In Vitro*," *Fertil Steril* 22:255 (1971).

Thibault, C. "*In Vitro* Fertilization of the Mammalian Egg," in *Fertilization*, Vol. 2, C. B. Metz and A. Monroy, Eds. (New York: Academic Press, 1969), p. 405.

SEXUALITY

Becker, R. F., J. A. Gehweiler and J. W. Wilson. *The Anatomical Basis of Medical Practice* (Baltimore: Williams & Wilkins, 1971).

Comfort, A., J. Wilson and N. Morris. *Man and Woman: The Encyclopedia of Adult Relationships* (New York: Greystone Press, 1974).

Diamond, M. "Sexual Anatomy and Physiology: Clinical Aspects," in *Human Sexuality: A Health Practitioner's Text*, R. Green, Ed. (Baltimore: Williams & Wilkins, 1974), pp. 20-34.

Diamond, M., A. L. Diamond and M. Mast. "Visual Sensitivity and Sexual Arousal Levels During the Menstrual Cycle," *J. Nerv. Ment. Dis.* 155: 170-176 (1972).

Green, R. "Taking a Sexual History," in *Human Sexuality: A Health Practitioner's Text*, R. Green, Ed. (Baltimore: Williams & Wilkins, 1975), pp. 9-19.

Masters, W. H. and V. E. Johnson. *Human Sexual Inadequacy* (Boston: Little, Brown & Co., 1970).

Oliven, J. F. *Clinical Sexuality*, 3rd ed. (Philadelphia: J. B. Lippincott Co., 1974).

Parkes, A. S. *Patterns of Sexuality and Reproduction* (New York: Oxford University Press, 1976).

Stenchever, M. S. *Human Sexual Behavior: A Workbook in Reproductive Biology* (Cleveland: Case Western Reserve University Press, 1974), distributed by Year Book Medical Publishers, Inc., Chicago.

Wolfers, H. "Psychological Aspects of Vasectomy," *Brit. Med. J.* 4:297 (1970).

REPRODUCTIVE FAILURE

Glezerman, M., B. Lunenfeld, G. Potashnik, G. Oelsner and R. Beer. "Retrograde Ejaculation: Pathophysiologic Aspects and Report of Two Successfully Treated Cases," *Fertil. Steril.* 27:796 (1976).

REPRODUCTIVE SENESCENCE

Bishop, M. W. H. "Aging and Reproduction in the Male," *J. Reprod. Fertil.* Suppl. 3:313-330 (1970).

Harbitz, T. B. "Testis Weight and the Histology of the Prostate in Elderly Men. Analysis in an Autopsy Series," *Acta. Pathol. Microbiol. Scand.* 81A:145-158 (1973).

Jones, E. C. "The Aging Ovary and Its Influence on Reproductive Capacity," *J. Reprod. Fertil.* Suppl. 12:17-30 (1970).

Judd, H. L., G. E. Judd, W. E. Lucas and S. S. C. Yen. "Endocrine Function of the Postmenopausal Ovary: Concentration of Androgens and Estrogens in Ovarian and Peripheral Vein Blood," *J. Clin. Endocrinol.* 39:1020-1024 (1974).

Kistner, R. W. "The Menopause," *Clin. Obstet. Gynecol.* 16:106-219 (1973).

Lang, W. R. and G. E. Aponte. "Gross and Microscopic Anatomy of the Aged Female Reproductive Organs," *Clin. Obstet. Gynecol.* 10:454-465 (1967).

Talbert, G. B. "Aging of the Reproductive System," in *Handbook of the Biology of Aging*, C. E. Finch and L. Hayflick, Eds. (New York: Van Nostrand Reinhold, 1976).

FERTILITY REGULATION

de Krester, D. "The Regulation of Male Fertility. The State of the Art and Future and Possibilities," *Contraception* 5:561 (1974).

Diczfalusy, E. "Current Developments, An Evaluation of Mode of Action of Contraceptive Drugs," *Am. J. Obstet. Gynecol.*, 100:136 (1968).

Flynn, A. M. "The Value of a Cervical Score in the Assessment of Ovarian Function," *J. Obstet. Gynaecol. Brit. Commonwlth.* 80(2):152-159 (1973).

Hubbard, C. W. *Family Planning: Education, Parenthood and Social Disease Control* (St. Louis, Missouri: C. F. Mosby, 1973).

Maas, J. M. "The Oral Contraceptives: Their Composition, Properties, Selection and Performance," *Clin. Med.* 7:14 (1974).

"Teaching of Human Reproduction and Training in Family Planning at Medical Schools," *IPPF Med. Bull.* 5(b) (1971).

The following periodicals should be useful for those interested in contraception or demography:

Advances in Planned Parenthood, Excerpta Medica Foundation, Amsterdam.

Contraception 13 (1977).

Demography 14 (1977).

Studies in Family Planning, Population Council, New York.

GLOSSARY

Adolescence: period of growth and development from puberty to somatic and psychosexual maturity; associated with the appearance of secondary sexual characteristics.

Agenesis: partial or complete failure of an organ or part of the body to form or develop to normal size and function.

Amastia: absence of mammary glands.

Amenorrhea: absence or abnormal cessation of menstrual bleeding.

Anaphrodisiac: drug or other agent (such as opiates) that abolishes sexual desire.

Anastomosis: formation of a passage, by surgery or disease, between two normally distinct spaces or organs.

Andrology: science dealing with fertility, infertility and diseases in the male; comparable to gynecology in the female.

Anomaly: remarkable anatomical, morphological, or physiological deviation from standard norms.

Anorexia: loss of appetite for food.

Anorexia nervosa: serious nervous syndrome due to loss of appetite, associated with excessive emaciation and hysterical aversion to food.

Aphrodisiac: drug or substance that stimulates sexual desire.

Areola (plural areolae): the dark pigmented skin immediately surrounding the nipple in both women and men.

Aspermia: failure of sperm formation in the testis, with resultant infertility.

Asthenozoospermia: (asthenospermia) decreased motility of sperm.

Atresia: imperforation or closure of a normal opening or canal; blighting and death of an ovarian follicle.

Autoerotism: sexual self-arousal without the participation of another person or presence of love object.

Azoospermia: absence of sperm in ejaculated semen.

Bisexuality: (AC-DC person) sexual attraction for and contact with members of both sexes.

Blastocyst: fertilized egg developed beyond the morula stage by the acquisition of a fluid-filled central cavity, the blastocele; a later stage in the development of the embryo which contains an inner cell mass, a trophoblast and a blastocele.

Buccal smear: a smear of cells lightly scraped from inside the mouth and prepared on a microscope slide to disclose the presence or absence of a sex chromatin spot in the nucleus of the cells; the spot is typically present (positive) in cells from females, but absent (negative) in cells from males.

Carcinoma: cancer.

Caudal: denoting a position or direction away from the head and more toward the tail.

Cephalic: pertaining to the head, or end of the body; the opposite of caudal.

Chorion: one of the placental membrane's peripheral parts of the conceptus, trophoblastic in origin and consisting of peripheral syncytiotrophoblast, underlying cytotrophoblast and central mesoblast.

Chromosome: one of 46 different-sized microscopic structures in the nucleus of each cell in the body on which are arranged, like beads on a string, the genes which carry an individual's genetic traits.

Cleavage: division of fertilized egg into equal cell halves, *i.e.*, 2-4-8-16-32-64, etc.; usually occurs in the oviduct; size of embryo remains unchanged, while cleavage cells (blastomeres) become smaller with each division.

Climacteric: a "critical" period of life occurring at the menopause of women, or at diminution of sexual activity in men; associated with endocrine, somatic and psychic changes.

Clitoris: the small, hooded organ in the female which is the counterpart of the penis in the male, located at the anterior part of the vulva, and used for sexual stimulation.

Clitorism: continuous erection or overdevelopment of the clitoris.

Coitus: sexual intercourse.

Coitus interfemora: insertion of the penis between the pressed thighs of the partner.

Conceptus: total components of the fertilized ovum and all of its derivatives from fertilization until birth, consisting of fetus and trophoblast.

Congenital: trait existing at or previous to birth.

Corpora cavernosa: pair of "cavernous" or spongy columns of erectile tissue in the penis below which is the urethra and corpus spongiosum.

Corpus luteum: endocrine gland formed in the ovary at the site of a ruptured Graafian follicle and release of egg; secretes progesterone; if conception and implantation take place, the corpus luteum increases in size and persists for several months; otherwise, it shrinks and degenerates.

Corpus spongiosum: erectile, spongy tissue of the penis surrounding the urethra.

Cortisol: naturally secreted glucocorticoid hormone of the adrenal cortex.

Cryptorchidism (cryptorchism): a developmental defect characterized by failure of one (unilateral) or two (bilateral) testes to descend into the scrotum; thus, they are retained in the abdomen or inguinal canal causing inhibition of spermatogenesis.

Cunnilingus: erotic stimulation of the female sex organs with the tongue and mouth, either in lesbian or normal heterosexual relations.

Cyproterone: a synthetic, hormonal, steroid substance, related to progesterone, which is highly potent as an antiandrogen.

Cyst: normal or abnormal sac containing fluid or other semisolid material with a distinct wall.

Cystocele: hernial protrusion of the urinary bladder through the vaginal wall; posterior wall of the bladder which, due to improper support, falls downward, pushing the anterior wall of the vagina before it.

Cytogenetics: branch of the science of heredity which deals with the chromosomes and genes, carriers of the genetic code, within the cell nucleus.

Dartos: tunica dartos, a layer of tissue beneath the skin of the scrotum.

Decidua: mucous part of the uterus which is shed at birth and menstruation; maternal part of placenta.

Desquamation: shedding of large sheets of epithelial cells as in vagina and skin.

Detumescence: subsiding of genital engorgement in penis or clitoris.

Diencephalon: posterior part of prosencephalon (anterior portion of brain) consisting of hypothalamus, thalamus, metathalamus and epithalamus.

Dildo: prosthetic penis, or an artificial instrument resembling a penis for use in sexual intercourse.

Diplegia: bilateral paralysis affecting similar parts of both sides of the body, *e.g.*, infantile, facial, masticatory, or spastic.

Distal: remote, farthest away.

Dorsal: pertaining to the back.

Dysgenesis: (gonadal) defective development in the presence of germinal cells.

Dysmenorrhea: pain associated with menstrual periods.

Dyspareunia: difficulty or emotional pain experienced by women during sexual intercourse as a result of cystitis, urethritis, or other organic disturbances.

Ectopic: located in an abnormal place, *e.g.*, an ectopic pregnancy is one in which the blastocyst is implanted outside the uterus, *e.g.*, in the oviduct, abdominal cavity, or some ligaments.

Endocrine: any of the ductless internal glands which secrete directly into the bloodstream or lymph system; pertaining to the ductless or hormone-secreting glands.

Endometriosis: where endometrial tissue is found outside the uterus or within the muscle of the uterine wall.

Endometrium: mucous lining of the uterus made of epithelium and stroma that is shed each month during menstrual bleeding; the thickness and structure of the endometrium vary with the phase of menstrual cycle.

Enteritis: inflammation of small intestines.

Enzyme: one of a class of organic compounds found in living organisms, which speed up or catalyze many of the metabolic processes.

Epididymis: cord-like duct which lies posterior to testis and is connected to it by efferent ductules.

Epididymitis: inflammation of epididymis.

Epimenorrhea: intermenstrual bleeding; uterine bleeding apart from normal menstruation.

Epiphyseal: pertaining to the peiphyses, the small bony centers where growth takes place at the ends of the long bones.

Epithelium: internal or external cell layer(s) covering a free surface or lumen, *e.g.*, skin, oviductal or vaginal surface.

Erectile: tissue capable of enlargement, *e.g.*, tissues of the penis, clitoris and mammary glands.

Estradiol: most active of naturally occurring estrogenic (steroid) hormone secreted chiefly by the ovarian follicular fluid, commercially prepared in various compounds, *e.g.*, estradiol benzoate and ethinyl estradiol.

Estrogen: natural or synthetic steroid hormone which can induce estrus in animals and cause estrogenic effect in women.

Etiology: theory, causation and origin of disease or problem.

Eunuch: a man castrated before puberty with secondary sexual characteristics similar to that of a woman.

Eunuchoid: possessing the sexual characteristics of a eunuch without actually being castrated.

Eunuchoidism: deficiency of testes or testicular secretion, with impaired sexual desire and eunuch-like symptoms, *e.g.*, castrated male.

Exhibitionism: psychosexual disorder of displaying oneself sexually, unexpectedly, to a stranger; patient is dependent for sexual arousal and orgasm on first having exposed his penis to a stranger.

Exudate: fluid, cells and cellular debris which have escaped from blood vessels and have been deposited in tissues or on tissue surface as a result of inflammation.

Fellatio: use of the mouth or tongue to stimulate penile erection.

Fetishism: psychosexual disorder in which sexual arousal is dependent upon a special object, *e.g.*, a shoe, an article of clothing.

Fibroid: benign tumor of the uterus, composed of muscle and fibrous tissue, *e.g.*, fibroma or myoma.

Follicle: secretory or excretory gland or sac-like structure.

Frigidity: lack of female sexual response or feeling.

FSH: (follicle-stimulating hormone), a hormone secreted by the anterior lobe of the pituitary gland which stimulates the production of estrogen.

Galactorrhea: excessive or spontaneous flow of milk from the female mammary gland as a result of endocrine or neuroendocrine imbalance.

Gamete: male or female reproductive cell capable of uniting in the process of fertilization to form a zygote; general term for sperm and egg.

Gonad: gland or organ producing gametes; general term for ovary or testis.

Gonadal agenesis: failure of the gonads to form or develop, *e.g.*, Turner's syndrome.

Gonadotropin: one of the hormones released by the anterior lobe of the pituitary gland, which regulates the functions of the ovary in the female and of the testis in the male.

Gravida: refers to a pregnant female; a woman with a first pregnancy is called primigravida or gravida I; with the second pregnancy, gravida II; and with the third, gravida III.

Gynecomastia: female-like development of mammary gland in men, naturally or as a result of hormonal treatment.

Haploid: a single representation of a chromosomal part in the cell; *e.g.*, sperm and eggs are haploid cells (all other cells in the body are diploid).

Hermaphrodite: a person with both female and male reproductive systems, usually with one sex dominating.

Hermaphroditism: congenital ambiguity of the reproductive structures so that the sex of the individual is not clearly defined as exclusively male or exclusively female; named for Hermes and Aphrodite, the Greek god and goddess of love.

Heterosexuality: innate sexual attraction toward a person of the opposite sex.

Hirsutism: abnormal presence of excessive hair, especially in females.

Homologous chromosomes: a pair of chromosomes carrying the same loci, one chromosome being derived from each parent. All 44 human autosomes are present as 22 homologous pairs of chromosomes. In women, the two X chromosomes are homologous. In men, the Y chromosome is a very small chromosome and only a small portion of its partner X chromosome has homologous genetic components.

Homosexuality: sexual attraction and contact between two persons of the same sex.

Hormone: chemical substances, produced by the endocrine glands and transported in the blood stream, to reach various receptor cells to regulate their growth and function.

Hydrocele of tunica vaginalis: accumulation of fluid within the process of tunica vaginalis, surrounding the testis, or within the spermatic cord.

Hymen: membrane which partly covers the entrance to the vagina and is stretched or broken during first intercourse.

Hyperkeratosis: hypertrophy of corneous layer of the skin.

Hyperplasia: abnormal multiplication or increase in the number of normal cells in normal arrangement in a tissue, causing an increase in size of tissue or organ.

Hyperspermia: semen ejaculate with volume greater than 6 ml.

Hypertrophy: abnormal enlargement or overgrowth of an organ or tissue due to increased size of its constituent cell.

Hypophysis: pituitary gland.

Hypoplasia: defective or incomplete development of a tissue or organ.

Hypospermia: semen ejaculate with a volume less than 2 ml.

Hypothalamus: the portion of the diencephalon which forms the floor and part of the lateral wall of the third ventricle in the brain, which releases neurohumoral substances, hypothalamic-releasing hormones, which regulate the secretory function of the pituitary gland.

Hysterectomy: surgical removal of part or all of the uterus and often with removal of one or both ovaries. Total hysterectomy of the uterus involves excision of the entire uterus, including the cervix, whereas in subtotal hysterectomy the cervix is left intact. Partial hysterectomy involves removal of only the cervix. In Wertheim's operation, ovaries and adjacent lymph nodes are removed during radical hysterectomy.

Idiopathic: self-originating, with unknown cause.

Impotence: inability to achieve penile erection.

Inguinal: in the region of the groin.

Inner cell mass: cells of the blastocyst that will develop into the body of the embryo and fetus.

Interstitial: within a tissue.

Introitus: the opening into the vagina.

Involution: degeneration of tissues or organs; the reverse of evolution of development, *e.g.*, involution of uterus and mammary gland.

In vitro: "in glass," or in a test tube or similar container in the laboratory.

In vivo: within the living body.

Ischemia: deficiency of blood supply in an organ due to obstruction or functional constriction of blood capillaries.

IUD: an intrauterine device inserted into the uterus to prevent conception.

Karotype: the visual display of a cell's chromosomes, photographed, measured, labeled and arranged according to international standards.

Kinocilia: motile cilia found primarily in the respiratory and urogenital tract.

Klinefelter's syndrome: a syndrome characterized by gynecomastia, increased urinary gonadotropins and hyalinization and fibrosis of the seminiferous tubules without involvement of the Leydig cells; patients usually have testicular dysfunction and extra X chromosome.

Labia majora: the larger or outer lips of the female external genitalia.

Labia minora: the smaller or inner lips of the female external genitalia.

Labium: a fleshy border or lip-shaped structure, *e.g.*, labium majus and labium minus of the female external genitalia.

Laparotomy: surgery in which the abdomen is opened to explore some internal organs.

Lesbian: a female homosexual.

Leukocyte: (white blood cell) a colorless, ameboid cell of the blood with a nucleus and cytoplasm.

Leydig cells: androgen-secreting cells found in the interstitial spaces between the seminiferous tubules of the testes.

LH (Luteinizing hormone): a hormone secreted by anterior pituitary gland; it induces release of the egg from the Graafian follicle and transforms the latter into a corpus luteum; also stimulates testosterone production in the male.

Lithotomy position: surgical position originally named for its use in cutting out bladder stones; the patient lies on the back, thighs spread, legs outstretched and feet lifted so that the genitals are easily examined.

Lumen: the cavity or channel within a tube or tubular organ, as in oviduct, uterus or vas deferens.

Luteal: pertaining to the corpus luteum, the stage in the menstrual cycle after ovulation, up to menstruation.

Mastectomy: surgical removal of the breasts.

Mastitis: inflammation of the mammary gland, characterized by a lumpy or nodular consistency.

Menarche: the establishment or beginning of menstrual function at puberty.

Menorrhagia: profuse or excessive uterine menstrual bleeding occurring at regular intervals, the period of menstrual flow being greater than the usual duration.

Metabolite: a chemical substance which appears in the body as a metabolic residue after the body has made use of or otherwise changed the parent compound.

Metastasis: a new growth of cancer, originated by cells from an older growth.

Metorrhagia: irregular or continuous bleeding from the uterus.

Miosis (meiosis): special cell division in the ovary or testes leading to a reduction of the chromosome number to one-half (haploid) of its original number (diploid) and to the formation of germ cells such as eggs or sperm.

Mons pubis: (mons veneris) fatty tissue over the female's pubic bone which becomes covered with hair at menarche.

Morula: fertilized egg which cleaves 8-80 (cells), forming a solid mass of cells called blastomeres.

Mullerian ducts: the structures in the fetus that will, in the female, develop into the uterus and oviducts.

Multipara (multiparous): woman who has given birth to two or more viable children, over a period of years or at one time. The term is often shortened to para II or para III. Primipara is used to denote a woman who has given birth to one viable child and nullipara or para 0 designates one who has never given birth to a viable child.

Myometrium: smooth muscle coat of the uterus, which forms the main mass of the organ.

Myotonia: increased muscle tension.

Necrosis: pathological death of a group of cells in contact with living cells.

Neuroendocrine: interaction of function between nerve cells of the central nervous system and hormone-secreting cells of the endocrine glands.

Normozoospermia (normospermia): semen ejaculate containing between 100-200 million sperm/ml.

Nullipara: a woman who has never given birth to a viable child.

Oligomenorrhea: abnormally infrequent or scanty menstruation.

Oligozoospermia (oligospermia): semen ejaculate containing less than 40 million sperm/ml.

Oocyte: a growing or full grown mammalian egg before it completes its maturational processes.

Oogenesis: formation of ova or egg cells.

Oogonium: primordial egg before it enters the prophase of miosis.

Oophorectomy: surgical removal of one or two ovaries; castration of the female.

Orchidectomy: removal of the testis; castration of the male.

Orgasmic platform: vasocongestion and contraction of the outer third of the vagina during plateau phase of sexual response.

Osteoporosis: abnormal porousness of bone by enlargement of its canals or the formation of abnormal spaces in response to failure of osteoblasta to lay down bone matrix.

Ostium: an opening into a tubular organ or between two distinct body cavities, *e.g.*, ostium of oviduct.

Ovariectomy: surgical removal of the ovary; castration of the female.

Ovotestis: an abnormally developed sex gland of which the cell structure partially resembles an ovary and partially a testis.

Ovulation: release of mature, unfertilized egg from the Graafian follicle of the ovary.

Ovum: female germ cell; egg cell.

Paraplegia: paralysis of lower parts of the body associated with uncoordinated mobility and affected sensation; caused by locomotor ataxia, anemia, brain lesions, injury or disease of the spinal cord, or chronic alcoholism.

Pathological: characterized by defect, disease or disorder of function.

Perineal: anatomical region between the thighs; in the male it is between the scrotum and the anus, in the female, between the mons veneris and the anus.

Peritoneum: the serous membrane lining the abdominopelvic walls and investing the viscera; it is a strong, colorless membrane with a smooth surface.

Placenta: union of conceptal and maternal parts which serve physiological exchange of nutrients and metabolic by-products between fetus and mother.

Plasma: the clear-fluid portion of the blood in which the red and white corpuscles are suspended.

Plexus: closely woven network or tangle of veins, nerves or lymphatic vessels.

Polymenorrhea: frequent and often profuse menstruation.

Polyzoospermia (polyspermia): semen ejaculate containing more than 250 million sperm/ml.

Pornography: photographs or written matter portraying sexual activity used for sexual stimulation.

Posterior fornix: the vault of the vagina at its posterior wall.

Postmenopausal: period of a woman's life after cessation of menstruation, usually after 50 years of age.

Pregnanediol: a hormonally inactive waste product mainly of progesterone, secreted and measured in the urine.

Premenopausal: period of a woman's life when menstruation becomes scanty or less frequent, usually between 45 and 50 years of age.

Prepuce: a fold of skin overlying the glans penis and the clitoris.

Priapism: persistent penile erection.

Primipara: a woman who has given birth to only one viable child.

Progesterone: one of the two sex hormones chiefly produced by the ovary in the corpus luteum, following ovulation.

Prolapse: downward displacement of herniation of an organ due to weakness of the supporting structures, *e.g.*, of vagina, uterus, or rectum.

Promiscuity: indiscriminate sexual activity with several partners.

Prosthesis: an artificial limb or organ of the body.

Pseudocyesis: spurious or false pregnancy.

Pseudohermaphrodite: a male hermaphrodite (with two testes) or a female hermaphrodite (with two ovaries) by contrast with a true hermaphrodite (with both testicular and ovarian tissue).

Rectocele: hernial protrusion of part of the rectum into the vagina causing the anterior wall of the rectum and the posterior wall of the vagina face down.

Retroversion: tipping of an entire organ backward; *e.g.*, turning backward of the entire uterus in relation to the pelvic axis.

Sadism: sexual satisfaction accomplished by inflicting pain and humiliation on partner.

Scopophilia: sexual satisfaction accomplished by viewing sexual acts or genitals of others.

Semen: thick, whitish, ejaculatory secretion of the male's reproductive organs; composed of sperm and seminal plasma; secreted by the prostate, seminal vesicles and various other male accessory glands.

Sertoli cells: elongated cells in the tubules of the testes, in which germinal epithelial cells are embedded during maturation.

Sex chromatin: sex-indicator spot normally in the nucleus of cells taken from the blood, skin or other tissues of females, but not males, as in buccal smear.

Sex flush: sexual arousal associated with superficial vasocongestive skin during the plateau phase of intercourse.

Sixty-nine: simultaneous fellatio and cunnilingus in which the partners are in a head-to-foot position of a sexual act.

Spermatocele: a cyst containing sperm and lying just above and posterior to the testis.

Smegma: whitish secretion accumulating under the prepuce of penis or clitoris.

Spermatogenesis: the process of formation of sperm including spermatocytogenesis and spermiogenesis.

Sphincter: ring-like band of muscle surrounding an opening which is opened or closed by muscular contraction.

Spinnbarkeit: formation of thread of cervical mucus when drawn over a glass slide.

Squamous: scaly or plate-like cell arrangement.

Teratozoospermia (tetratospermia): semen ejaculate containing more than 40% abnormal sperm.

Tetraplegia: paralysis of four extremities.

Thromboembolism: obstruction of a blood vessel with a thrombus (clot) released from its site of formation.

Transsexualism: sexual deviation in which a person desires to change his or her sex, requiring surgical manipulation, intensive hormone therapy, psychotherapy and legal sex reassignment; conviction and compulsion to have the body, appearance and social status of the opposite sex.

Transudate: a fluid which has passed from serum or tissue through mucous membrane, as a result of sexual excitement or inflammation.

Transvestitism: sexual deviation in which a person dresses in the clothing of the opposite sex; condition where one feels compelled to cross-dress, often in relation to sexual arousal and attainment of orgasm.

Troilism: homosexual and heterosexual activity involving three persons, one of them of an opposite sex.

Trophoblast: peripheral layer of cells of the blastocyst, the derivatives of the fertilized egg which become the placenta and the fetal membranes, trophoblast of the preimplanting embryo is comprised of about 100 cells.

Tunica: covering or coat of membrane.

Unilateral: affecting only one side of the body, or specific organ or point is located on only one side of the body.

Vaginismus: painful muscular spasms in the vaginal walls as a result of emotional problems; may be so severe as to interfere with intercourse.

Vaginitis: inflammation of the vaginal epithelium due to infectious organisms, venereal disease or fungus infections, *e.g., Candida albicans*; usually associated with painful irritation and purulent leukorrheal discharge.

Variococele: dilatation of pampiniform plexus about the testes where the veins passing near the vas deferens drain into the internal spermatic veins in the region of the internal inguinal ring.

Vas deferens: excretory duct of testes epididymis.

Vasectomy: sterilization procedure of men when the vas deferens are cut or tied off, eliminating the release of sperm without affecting potency and sexuality.

Virilization: development in a female of male secondary sexual characteristics, including enlargement of the clitoris, hirsutism, deepening of the voice, decreased breast size and baldness.

Voyeurism: sexual deviation in which a person experiences pleasure by observing sexual organs or sexual acts by others.

Vulva: the external part of the female sex organs.

Wolffian ducts: embryonic structures which develop into the internal reproductive anatomy of the male, attached to the testis.

Zygote: egg after fertilization, and after fusion of the female pronucleus with male pronucleus, but before first cleavage; cell resulting from fusion of two gametes; the fertilized ovum.

INDEX